SRA

Connecting Math Concepts

Level C Workbook 2

COMPREHENSIVE EDITION

A DIRECT INSTRUCTION PROGRAM

Bothell, WA • Chicago, IL • Columbus, OH • New York, NY

MHEonline.com

Copyright © 2012 The McGraw-Hill Companies, Inc.

All rights reserved. No part of this publication may be reproduced or distributed in any form or by any means, or stored in a database or retrieval system, without the prior written consent of The McGraw-Hill Companies, Inc., including, but not limited to, network storage or transmission, or broadcast for distance learning.

Send all inquiries to:
McGraw-Hill Education
4400 Easton Commons
Columbus, OH 43219

ISBN: 978-0-02-103577-9
MHID: 0-02-103577-6

Printed in the United States of America.

8 9 10 11 QLM 18 17 16 15

The McGraw·Hill Companies

Lesson 71

Name ____________________

Part 1

a.

4 in.

1 in. 1 in.

4 in.

__________ =

__________ = ____ in.

---

b.

5 cm

4 cm 4 cm

5 cm

__________ =

__________ = ____ cm

Part 2

a. 35 + _____ = 60

⟶ ______

b. 110 + _____ = 145

⟶ ______

c. 18 + _____ = 41

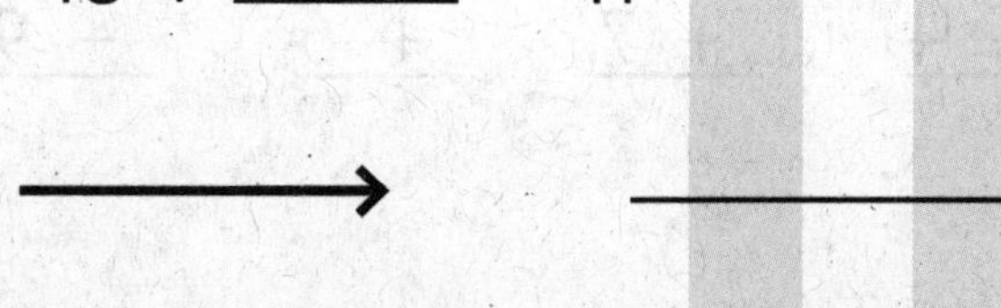

⟶ ______

d. 500 + _____ = 800

⟶ ______

Copyright © The McGraw-Hill Companies, Inc.

Lesson 71

Name ________________________

Part 3

a. one hundred twenty ______

b. thirty-eight ______

c. sixteen ______

d. sixty ______

e. one hundred thirty ______

f. two hundred six ______

g. thirty-four ______

Independent Work

Part 4

a. $54 + 9 =$ ______

b. $9 + 136 =$ ______

c. $12 + 9 =$ ______

d. $55 + 9 =$ ______

Part 5

a. $\begin{array}{r} 832 \\ -\ 423 \\ \hline \end{array}$

b. $\begin{array}{r} 328 \\ -\ 119 \\ \hline \end{array}$

c. $\begin{array}{r} 754 \\ -\ 547 \\ \hline \end{array}$

d. $\begin{array}{r} 38 \\ +\ 28 \\ \hline \end{array}$

Part 6

a. $R > T$
$T > V$

______

b. $15 < B$
$J < 15$

______

Part 7

a. $\begin{array}{r} 7 \\ +\ 9 \\ \hline \end{array}$

b. $\begin{array}{r} 3 \\ +\ 6 \\ \hline \end{array}$

c. $\begin{array}{r} 14 \\ -\ 10 \\ \hline \end{array}$

d. $\begin{array}{r} 5 \\ +\ 10 \\ \hline \end{array}$

e. $\begin{array}{r} 4 \\ +\ 4 \\ \hline \end{array}$

f. $\begin{array}{r} 7 \\ +\ 7 \\ \hline \end{array}$

g. $\begin{array}{r} 6 \\ +\ 6 \\ \hline \end{array}$

h. $\begin{array}{r} 9 \\ +\ 9 \\ \hline \end{array}$

Copyright © The McGraw-Hill Companies, Inc.

Lesson 72

Name ______________________

Part 1

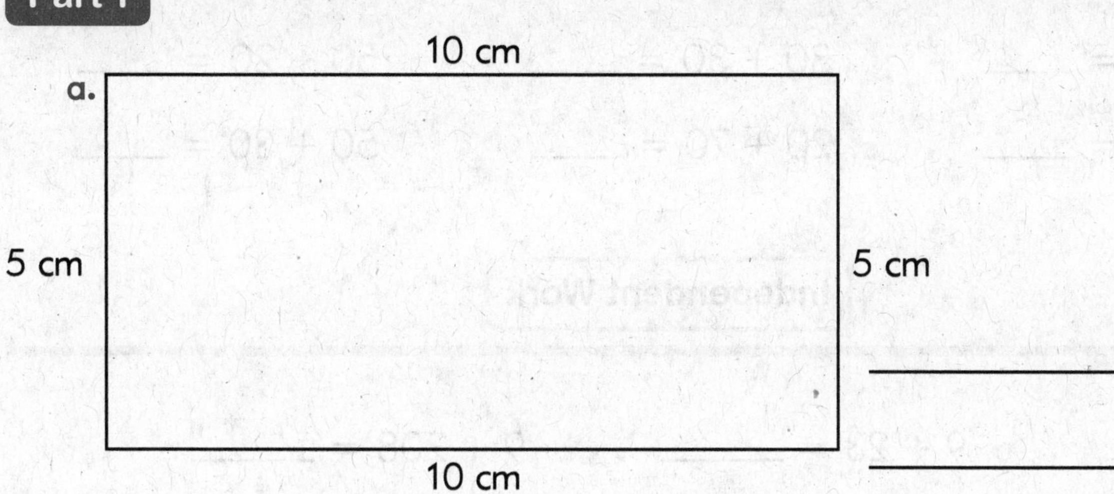

________ = ▭

________ = ▭
▭ cm

b.

1 in.

2 in. 2 in.

1 in.

________ = ▭

________ = ▭
▭ in.

Part 2

a. fifty-three ______

b. fourteen ______

c. one hundred eight ______

d. sixty ______

e. three hundred twelve ______

f. seventy-two ______

g. two hundred ______

Copyright © The McGraw-Hill Companies, Inc.

Lesson 72

Name ______________________

Part 3

a. $10 + 80 =$ ____ b. $30 + 30 =$ ____ c. $50 + 20 =$ ____

d. $10 + 40 =$ ____ e. $20 + 70 =$ ____ f. $50 + 30 =$ ____

Independent Work

Part 4

a. $9 + 23 =$ ______ b. $9 + 208 =$ ______

c. $9 + 63 =$ ______ d. $9 + 81 =$ ______

Part 5

a. $V > C$
$C > R$

____

b. $P < 50$
$T < P$

____

Part 6

a.	b.	c.	d.	e.
$\begin{array}{r} 3 \\ +\ 7 \\ \hline \end{array}$	$\begin{array}{r} 5 \\ +\ 5 \\ \hline \end{array}$	$\begin{array}{r} 9 \\ -\ 3 \\ \hline \end{array}$	$\begin{array}{r} 12 \\ -\ 6 \\ \hline \end{array}$	$\begin{array}{r} 12 \\ -\ 10 \\ \hline \end{array}$

f.	g.	h.	i.	j.
$\begin{array}{r} 4 \\ +\ 9 \\ \hline \end{array}$	$\begin{array}{r} 6 \\ -\ 3 \\ \hline \end{array}$	$\begin{array}{r} 10 \\ -\ 3 \\ \hline \end{array}$	$\begin{array}{r} 16 \\ -\ 10 \\ \hline \end{array}$	$\begin{array}{r} 3 \\ +\ 6 \\ \hline \end{array}$

k.	l.	m.	n.	o.
$\begin{array}{r} 8 \\ +\ 3 \\ \hline \end{array}$	$\begin{array}{r} 20 \\ -\ 10 \\ \hline \end{array}$	$\begin{array}{r} 5 \\ +\ 10 \\ \hline \end{array}$	$\begin{array}{r} 8 \\ -\ 4 \\ \hline \end{array}$	$\begin{array}{r} 3 \\ +\ 9 \\ \hline \end{array}$

Copyright © The McGraw-Hill Companies, Inc.

Lesson 73

Name ____________________

Part 1

a.

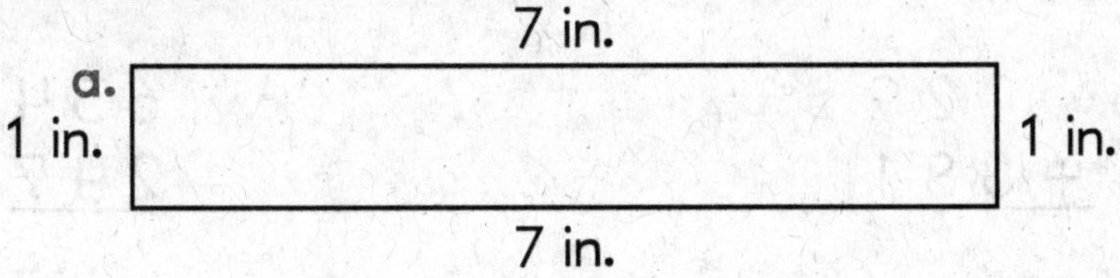

__________ = ▢

__________ = ▢
▢ in.

b.

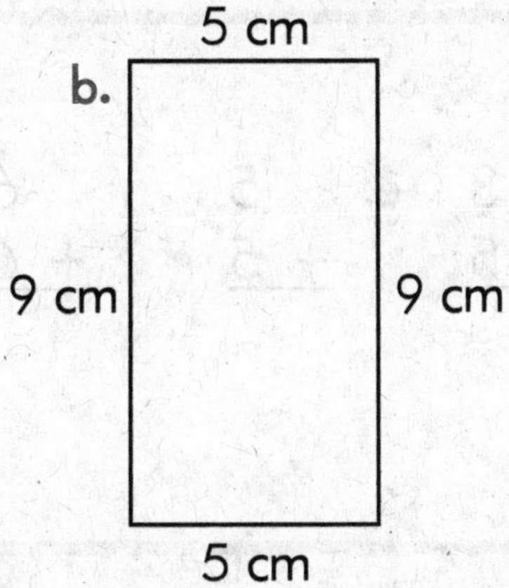

__________ = ▢

__________ = ▢
▢ cm

Part 2

a. 70 – 20 = ____ b. 70 + 20 = ____ c. 80 – 10 = ____

d. 40 – 30 = ____ e. 40 + 30 = ____ f. 30 + 30 = ____

Independent Work

Part 3

a. 19 – 10 = ____ b. 43 + 9 = ____ c. 48 + 10 = ____

Part 4

a. 37 + ▢ = ▢

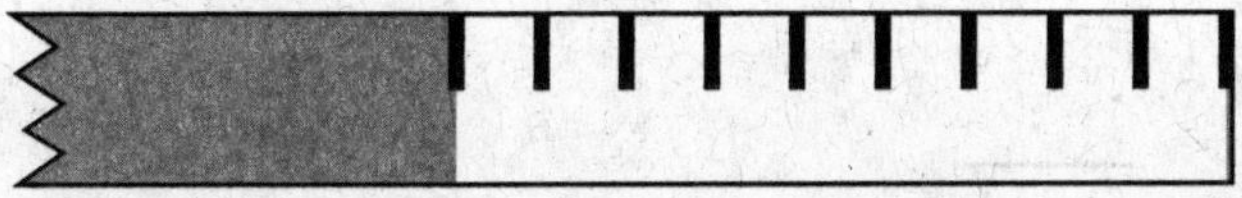

b. ▢ + ▢ = 52

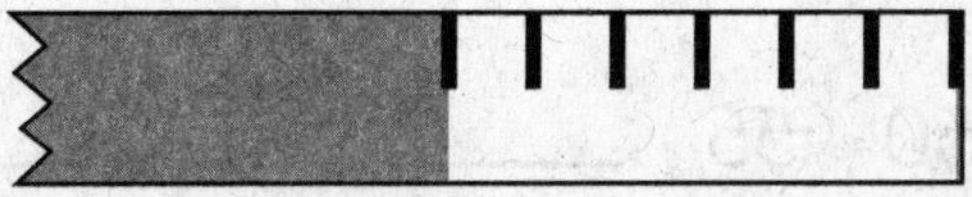

Copyright © The McGraw-Hill Companies, Inc.

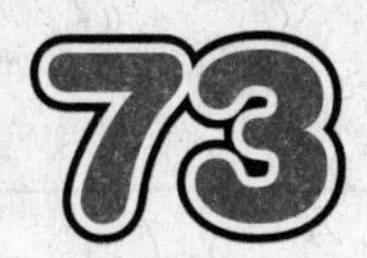

Lesson 73

Name ____________________

Part 5

a. $\begin{array}{r} 872 \\ -\ 456 \\ \hline \end{array}$

b. $\begin{array}{r} 29 \\ +\ 381 \\ \hline \end{array}$

c. $\begin{array}{r} 654 \\ -\ 247 \\ \hline \end{array}$

Part 6

a. $\begin{array}{r} 4 \\ +\ 3 \\ \hline \end{array}$
b. $\begin{array}{r} 10 \\ -\ 7 \\ \hline \end{array}$
c. $\begin{array}{r} 8 \\ +\ 8 \\ \hline \end{array}$
d. $\begin{array}{r} 3 \\ +\ 5 \\ \hline \end{array}$
e. $\begin{array}{r} 5 \\ +\ 5 \\ \hline \end{array}$
f. $\begin{array}{r} 6 \\ +\ 6 \\ \hline \end{array}$
g. $\begin{array}{r} 12 \\ -\ 3 \\ \hline \end{array}$
h. $\begin{array}{r} 16 \\ -\ 8 \\ \hline \end{array}$

Part 7 Write the sign >, <, or =.

a. 50 + 6 ☐ 55

b. 40 + 30 ☐ 69

c. 15 ☐ 5 x 3

d. 8 ☐ 2 x 5

Part 8 Write the missing numbers.

a. 18 20 ____ ____ ____ ____ ____

b. 50 55 ____ ____ ____ ____ ____

Copyright © The McGraw-Hill Companies, Inc.

Lesson 

Name ____________________

Part 1

a. **27** 20 30

b. **34** 30 40

c. **88** 80 90

d. **26** 20 30

e. **12** 10 20

f. **63** 60 70

g. **18** 10 20

Part 2

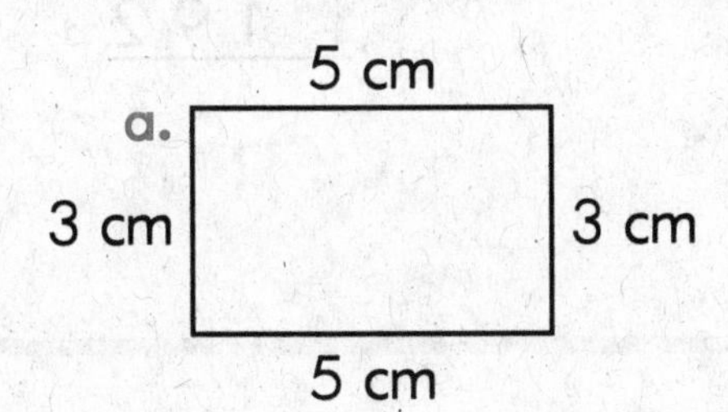

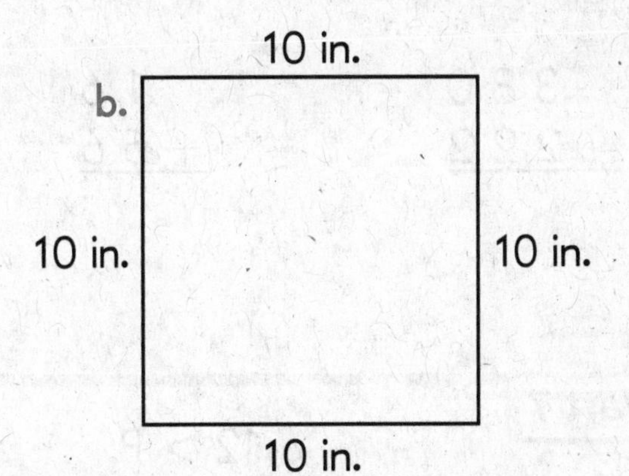

______ =

______ = ______ cm

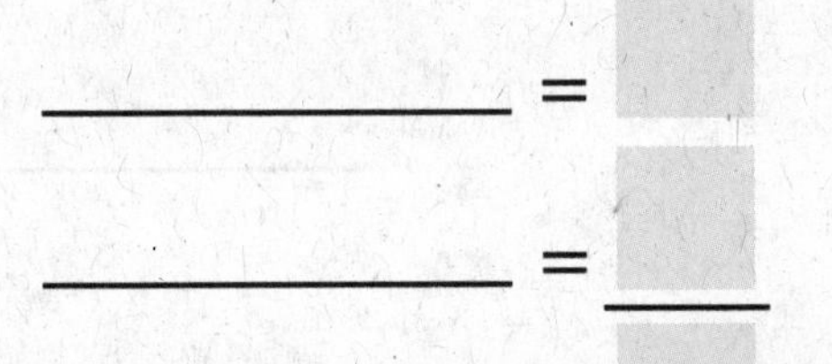

______ =

______ = ______ in.

Copyright © The McGraw-Hill Companies, Inc.

Part 3

a. 50 + 50 = ______ b. 90 + 40 = ______ c. 80 + 30 = ______

Part 4

a. 9 ⟶ 15

b. 9 ⟶ 12

c. 9 ⟶ 18

d. 9 ⟶ 11

e. 9 ⟶ 14

f. 9 ⟶ 17

Lesson 74

Name ______________________

Independent Work

Part 5

a. 29 + 9 = ____ b. 36 + 9 = ____ c. 15 + ____ = 25

d. 9 + 54 = ____ e. 30 + ____ = 38

Part 6

a.
$$\begin{array}{r} 380 \\ -\ 222 \\ \hline \end{array}$$

b.
$$\begin{array}{r} 16 \\ +\ 56 \\ \hline \end{array}$$

c.
$$\begin{array}{r} 29 \\ +\ 76 \\ \hline \end{array}$$

d.
$$\begin{array}{r} 582 \\ -\ 192 \\ \hline \end{array}$$

Part 7

a. $12 > P$
$P > J$

____

b. $160 < R$
$T < 160$

____

Part 8 Complete each place-value equation.

a. 100 + 0 + 5 = ______ b. 40 + 4 = ______

c. ________________ = 52 d. ________________ = 280

e. 500 + 0 + 0 = ______ f. ________________ = 800

Copyright © The McGraw-Hill Companies, Inc.

Lesson 75

Name ______________________

Part 1

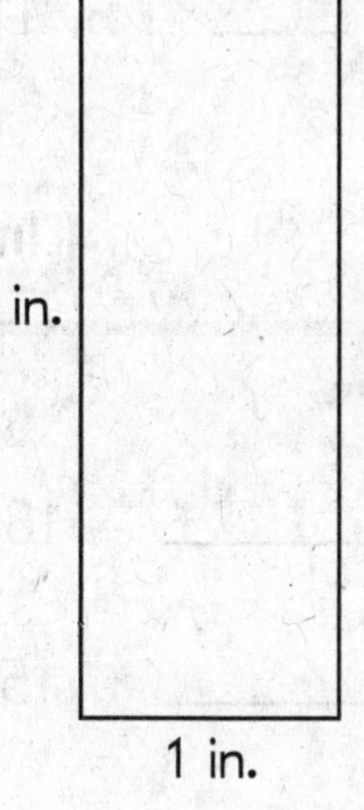

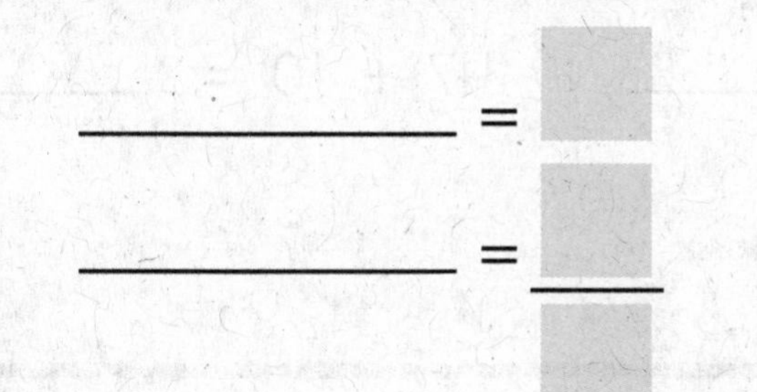

______________ = ▭

______________ = ▭/▭ ____

______________ = ▭

______________ = ▭/▭ ____

Part 2

a. 9 —→ 16

b. 9 —→ 13

c. 9 —→ 17

d. 9 —→ 15

e. 9 —→ 11

f. 9 —→ 14

Part 3

a. **59** 50 60

b. **74** 70 80

c. **76** 70 80

d. **16** 10 20

e. **92** 90 100

f. **14** 10 20

g. **26** 20 30

Copyright © The McGraw-Hill Companies, Inc.

Lesson 75

Name ____________________

Part 4

a. 160 – 150 = ____   b. 110 – 80 = ____   c. 120 – 100 = ____

Independent Work

Part 5

a. 56 – 10 = ______   e. 135 + 10 = ______   h. 195 – 10 = ______

b. 248 – 10 = ______   f. 153 – 10 = ______   i. 92 – 10 = ______

c. 140 + 10 = ______   g. 76 + 10 = ______   j. 47 + 10 = ______

d. 68 – 10 = ______

Part 6

a.	b.	c.	d.	e.	f.	g.	h.
9	4	15	9	2	8	18	5
+ 6	+ 9	– 6	+ 7	+ 9	+ 9	– 9	+ 9

i.	j.	k.	l.	m.	n.	o.	p.
17	9	9	1	13	9	19	6
– 8	+ 3	+ 9	+ 9	– 4	+ 7	– 10	+ 9

Part 7

a.	b.	c.
268	827	358
+ 402	– 363	+ 639

Copyright © The McGraw-Hill Companies, Inc.

Lesson 76

Name ____________________

Part 1

a. 9 ——→ 15

____

b. 9 ——→ 12

____

c. 9 ——→ 17

____

d. 9 ——→ 13

____

e. 9 ——→ 16

____

f. 9 ——→ 18

____

Part 2

a. $120 - 10 =$ ______

b. $160 - 10 =$ ______

c. $160 - 80 =$ ______

d. $90 + 40 =$ ______

e. $120 + 20 =$ ______

f. $10 + 170 =$ ______

Independent Work

Part 3

a.

5 in.

4 in.

____________ = ☐

____________ = ☐/☐ ______

b.

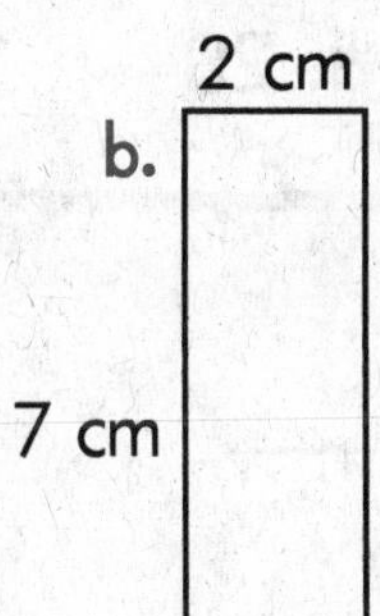

____________ = ☐

____________ = ☐/☐ ______

Copyright © The McGraw-Hill Companies, Inc.

Lesson 76

Name ______________________

Part 4

a. $6 + 9$ b. $10 + 9$ c. $9 + 5$ d. $9 + 7$ e. $18 - 9$ f. $12 - 3$ g. $4 + 9$ h. $15 - 6$

Part 5

a. $463 + 429$

b. $694 - 177$

c. $248 - 139$

Part 6

Circle the tens number that is closer.

a. **81** 80 90

b. **64** 60 70

c. **36** 30 40

d. **28** 20 30

e. **77** 70 80

f. **43** 40 50

g. **59** 50 60

Part 7

a. 68 + 9 = _____ b. 19 + 9 = _____ c. 45 + 9 = _____

d. 11 + 9 = _____ e. 73 + 9 = _____ f. 84 + 9 = _____

Copyright © The McGraw-Hill Companies, Inc.

Lesson 77

Name ______________________

Part 1

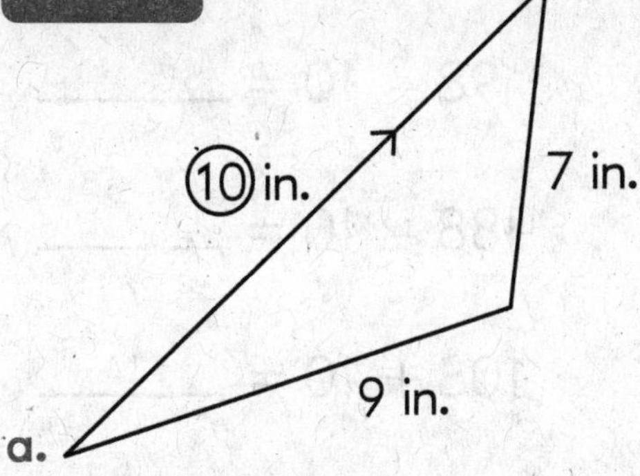

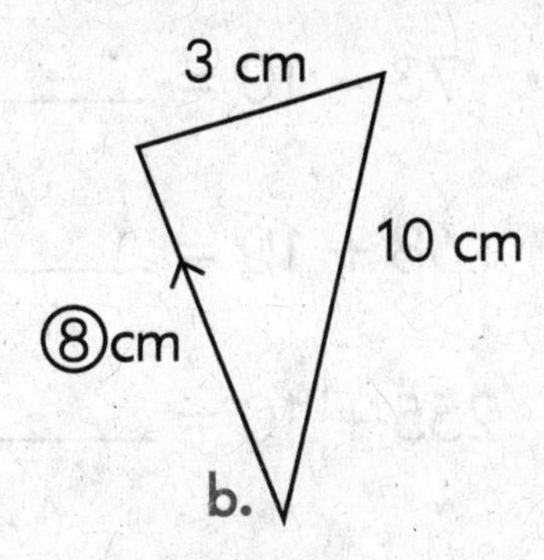

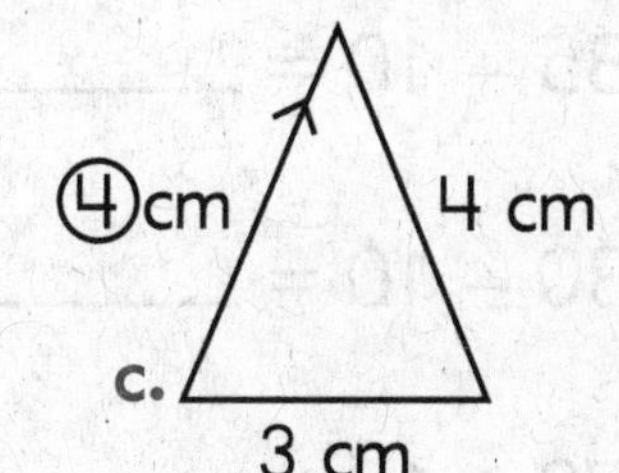

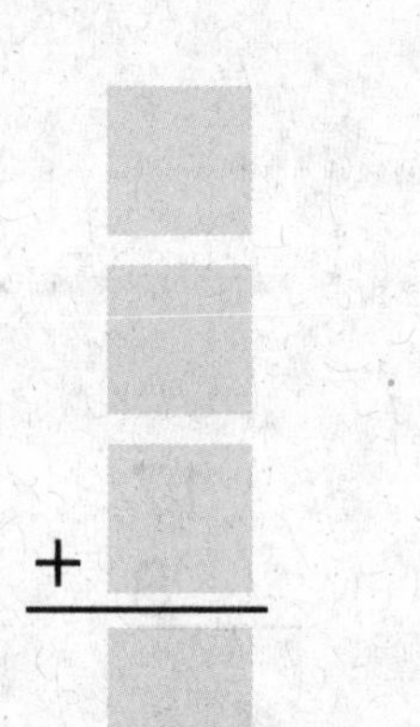

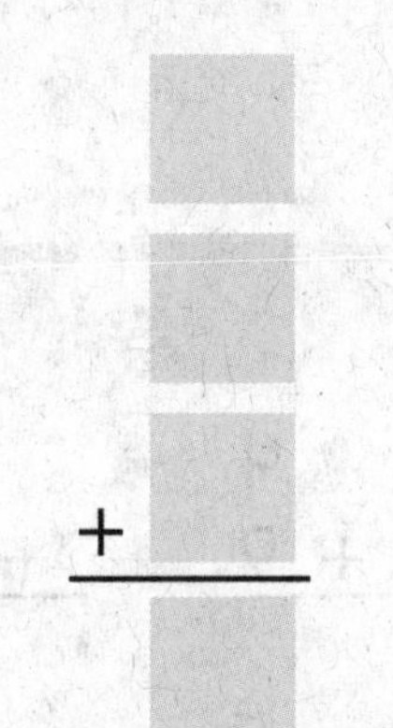

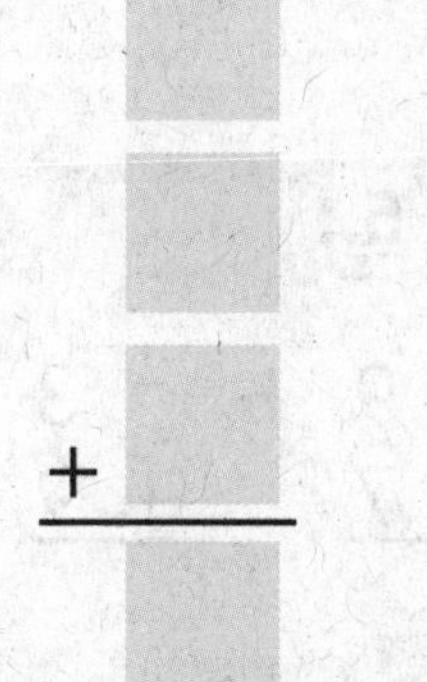

Part 2

a. $20 + 70 =$ ______ b. $120 - 90 =$ ______ c. $170 - 20 =$ ______

d. $30 + 100 =$ ______ e. $100 + 70 =$ ______ f. $100 - 80 =$ ______

Independent Work

Part 3

Write 2 subtraction facts.

a. 9 —— → 14

____

____

b. 9 —— → 11

____

____

c. 9 —— → 18

____

d. 9 —— → 15

____

____

e. 9 —— → 17

____

____

f. 9 —— → 13

____

____

Copyright © The McGraw-Hill Companies, Inc.

Lesson 77

Name ______________________

Part 4

a. 142 + 10 = ______

b. 85 − 10 = ______

c. 130 − 10 = ______

d. 73 + 10 = ______

e. 73 − 10 = ______

f. 68 + 10 = ______

g. 255 + 10 = ______

h. 98 − 10 = ______

i. 488 − 10 = ______

j. 108 + 10 = ______

Part 5

a.	b.	c.	d.	e.	f.	g.	h.
9	3	14	4	9	3	17	9
+ 7	+ 9	− 5	+ 9	+ 6	+ 9	− 8	+ 10

Part 6

Circle the tens number that is closer.

a. **68** 60 70

b. **33** 30 40

c. **88** 80 90

d. **24** 20 30

e. **19** 10 20

f. **44** 40 50

g. **52** 50 60

Part 7

a. 42 + 9 = ____

b. 18 + 9 = ____

c. 75 + 9 = ____

d. 56 + 9 = ____

e. 21 + 9 = ____

f. 37 + 9 = ____

Copyright © The McGraw-Hill Companies, Inc.

Lesson  Name ______________________

Part 1

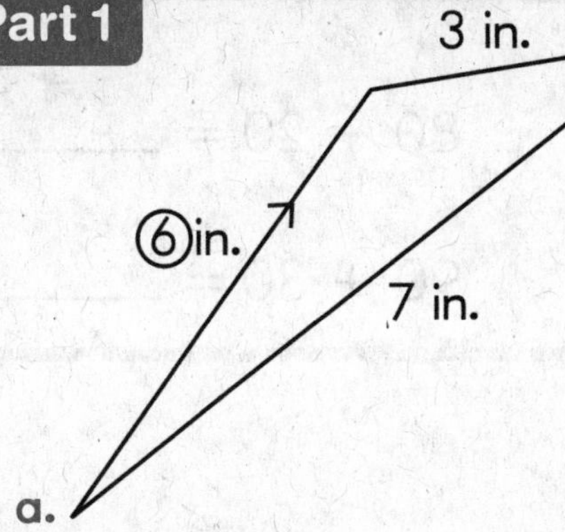

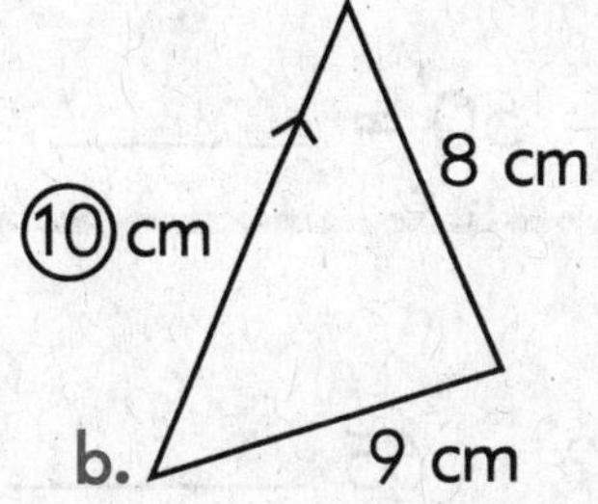

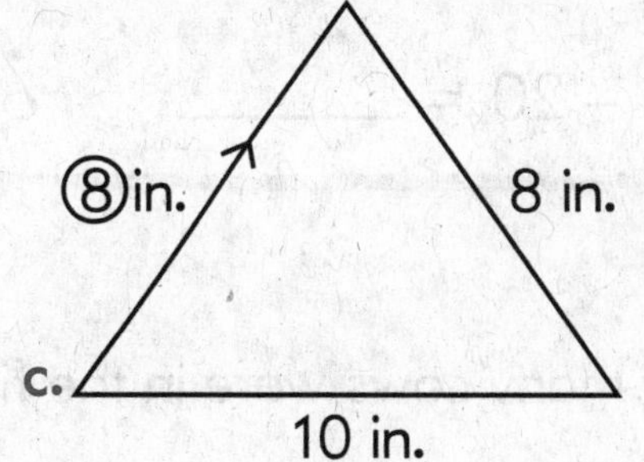

+ ____    + ____    + ____

Part 2

a. 270 – 10 = ______ b. 180 – 80 = ______ c. 40 + 30 = ______

d. 70 + 70 = ______ e. 130 – 30 = ______ f. 10 + 90 = ______

Independent Work

Part 3

a. 17 − 8   b. 17 − 7   c. 15 − 6   d. 15 − 5   e. 6 + 9   f. 3 + 9   g. 12 − 6   h. 9 + 5

Part 4

a. 176 + 316   b. 265 − 129   c. 346 − 237

Copyright © The McGraw-Hill Companies, Inc.

Lesson 

Name ____________________

Part 1

a. $180 - 90 =$ ______ b. $40 + 30 =$ ______ c. $80 + 20 =$ ______

d. $110 - 20 =$ ______ e. $60 - 50 =$ ______ f. $90 + 30 =$ ______

Part 2

a. How many cows were in the field? 65 ______________

b. How many feet long was the truck? 38 ______________

c. How many years older is Mary? 10 ______________

Part 3

a. $5 + 4 =$ ____	f. $14 - 10 =$ ____	k. $9 - 5 =$ ____	p. $13 - 9 =$ ____
b. $10 - 4 =$ ____	g. $8 - 4 =$ ____	l. $4 + 9 =$ ____	q. $10 - 6 =$ ____
c. $4 + 4 =$ ____	h. $4 + 6 =$ ____	m. $14 - 4 =$ ____	r. $4 + 10 =$ ____
d. $10 + 4 =$ ____	i. $4 + 5 =$ ____	n. $10 - 6 =$ ____	s. $9 - 4 =$ ____
e. $13 - 4 =$ ____	j. $9 + 4 =$ ____	o. $6 + 4 =$ ____	t. $5 + 4 =$ ____

Copyright © The McGraw-Hill Companies, Inc.

Independent Work

Part 4

a. $6 + 9$ b. $3 + 9$ c. $13 - 9$ d. $9 + 9$ e. $15 - 5$ f. $15 - 6$ g. $12 - 9$ h. $11 - 9$

Lesson 80

Name ______________________

Part 1

a. 90 + 90 = ______ b. 100 – 80 = ______ c. 120 – 60 = ______

d. 50 + 50 = ______ e. 90 + 20 = ______ f. 140 – 10 = ______

Part 2

a. How many houses did they paint? 18 ______________

b. How many pounds heavier is the black car? 258 ______________

c. How many flies were in the barn? 732 ______________

d. How many years younger is Jill? 40 ______________

e. How many leaves were on the tree? 932 ______________

Part 3

a. 14 – 10 = ____	f. 13 – 9 = ____	k. 5 + 4 = ____	p. 4 + 9 = ____
b. 4 + 6 = ____	g. 4 + 10 = ____	l. 10 – 4 = ____	q. 14 – 4 = ____
c. 8 – 4 = ____	h. 10 – 6 = ____	m. 4 + 4 = ____	r. 10 – 6 = ____
d. 9 + 4 = ____	i. 9 – 4 = ____	n. 13 – 4 = ____	s. 9 – 5 = ____
e. 4 + 5 = ____	j. 6 + 4 = ____	o. 10 + 4 = ____	t. 6 + 4 = ____

Copyright © The McGraw-Hill Companies, Inc.

Lesson 

Name ______________________

Independent Work

Part 4

a. $\begin{array}{r} 43 \\ +\ 9 \\ \hline \end{array}$

b. $\begin{array}{r} 146 \\ +\ 49 \\ \hline \end{array}$

c. $\begin{array}{r} 790 \\ -385 \\ \hline \end{array}$

d. $\begin{array}{r} 13 \\ 7 \\ +68 \\ \hline \end{array}$

e. $\begin{array}{r} 572 \\ +\ 18 \\ \hline \end{array}$

Part 5

a. $\begin{array}{r} 15 \\ -\ 9 \\ \hline \end{array}$

b. $\begin{array}{r} 18 \\ -\ 9 \\ \hline \end{array}$

c. $\begin{array}{r} 14 \\ -\ 9 \\ \hline \end{array}$

d. $\begin{array}{r} 12 \\ -\ 9 \\ \hline \end{array}$

Part 6

a. 42 + 10 = _____ b. 75 + 9 = _____ c. 41 + 9 = _____

d. 26 + 9 = _____ e. 83 + 10 = _____ f. 17 + 9 = _____

Part 7 Write the sign >, <, or =.

a. 52 + 10 ☐ 61 b. 65 + 9  74

c. _____ cents 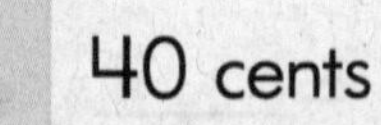 40 cents d. 38 cents  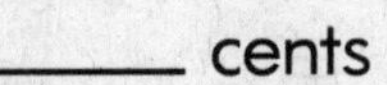 _____ cents

Copyright © The McGraw-Hill Companies, Inc.

Lesson 81

Name ________________________

Part 1

a. How many black cats did they have? 17 ________________

b. How many small boys were in the room? 32 ________________

c. How many big boxes are on the truck? 108 ________________

d. How many long snakes were there? 5 ________________

e. How many blue birds are in the tree? 21 ________________

Independent Work

Part 2

a. $100 - 50 =$ ______ b. $160 - 80 =$ ______ c. $40 + 90 =$ ______

d. $30 + 70 =$ ______ e. $90 + 70 =$ ______ f. $70 - 30 =$ ______

Part 3

a.	b.	c.	d.	e.	f.	g.	h.
16	15	12	15	17	12	14	18
− 9	− 9	− 6	− 5	− 9	− 9	− 9	− 9

Part 4

a. $B < P$
$P < K$
____

b. $12 < K$
$K < T$
____

c. $5 > T$
$J > 5$
____

Copyright © The McGraw-Hill Companies, Inc.

Lesson 82

Name ______________________________

Part 1

a. 36 − 12 = ____

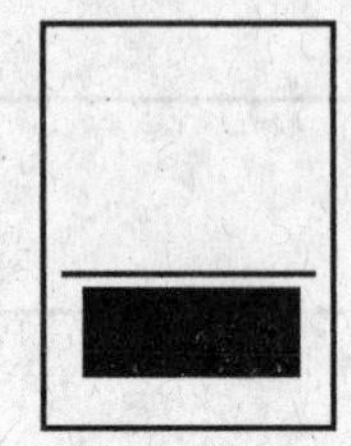

b. 81 − 48 = ____

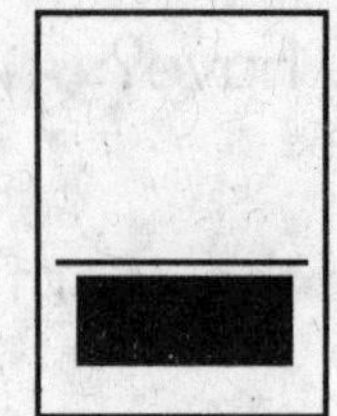

c. 17 + 41 = ____

Part 2

a. How many short pencils are there? 10 ____________

b. How many inches shorter is the fence? 216 ____________

c. How many white goats does Sally have? 25 ____________

d. How many broken eggs were there? 144 ____________

e. How many cars are in the lot? 93 ____________

Independent Work

Part 3

a. 70 + 20 = ______ b. 50 + 50 = ______ c. 180 − 90 = ______

Part 4

a. 9 + 26 b. 9 + 37 c. 580 + 212 d. 486 + 409 e. 309 + 184 f. 783 − 749

Part 5

a. 4 + 4 b. 9 + 9 c. 7 + 7 d. 5 + 5 e. 8 + 8 f. 6 + 6

Copyright © The McGraw-Hill Companies, Inc.

Lesson 

Name ______________________

Part 1

a. 59
+ 13

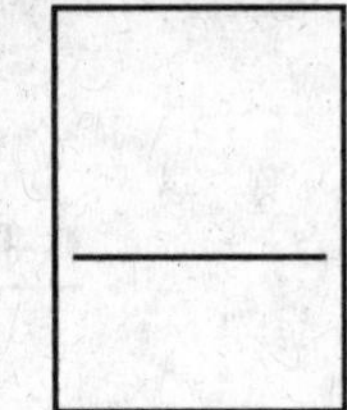

b. 42
+ 16

c. 68
− 19

Part 2

a. 427
+ 281

b. 690
+ 52

c. 162
+ 563

Independent Work

Part 3

a. 14
+ 9

b. 68
+ 18

c. 35
+ 35

d. 74
+ 24

e. 16
− 8

f. 18
− 13

g. 16
− 8

h. 19
− 9

i. 42
− 36

j. 17
− 8

Part 4

a. $56 < T$
$T < M$
____

b. $17 > B$
$B > 12$
____

c. $P > T$
$R > P$
____

Copyright © The McGraw-Hill Companies, Inc.

Lesson 84

Name ____________________

Part 1

a. 368 + 213

b. 159 + 450

c. 159 + 415

d. 792 + 83

Independent Work

Part 2

Work the estimation problems.

a. 51 − 18

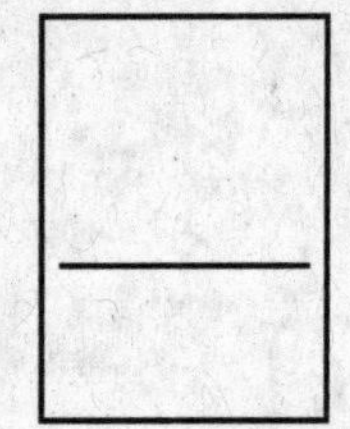

b. 47 + 42

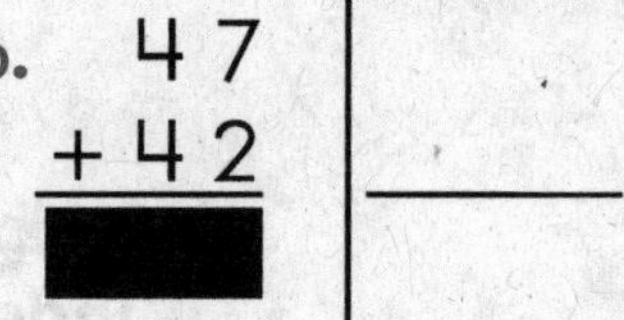

c. 36 + 29

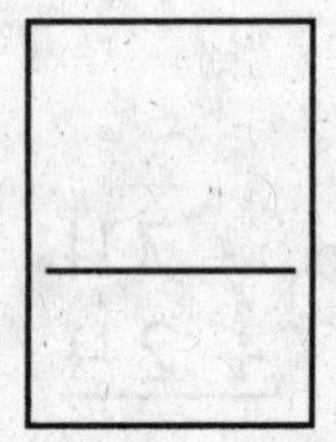

d. 92 − 71

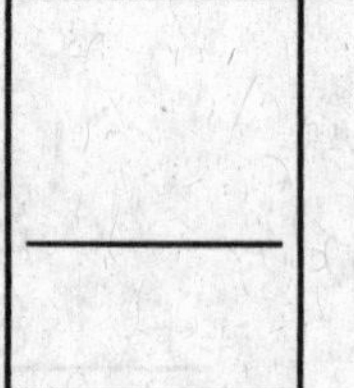

Part 3

a. 527 + 149

b. 368 + 328

c. 638 + 43

d. 498 − 349

Part 4

a. 19 − 9

b. 14 − 9

c. 12 − 9

d. 16 − 9

e. 15 − 9

Copyright © The McGraw-Hill Companies, Inc.

Lesson 

Name ____________________

Part 1

a. $\begin{array}{r} 182 \\ +\ 385 \\ \hline \end{array}$ b. $\begin{array}{r} 147 \\ +\ 23 \\ \hline \end{array}$ c. $\begin{array}{r} 496 \\ +\ 133 \\ \hline \end{array}$ d. $\begin{array}{r} 837 \\ +\ 147 \\ \hline \end{array}$

Part 2

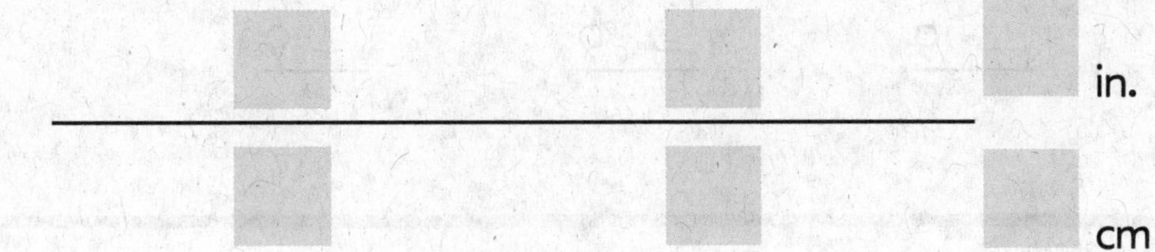

Independent Work

Part 3 Write the sign >, <, or =.

a. 50 + 20 ☐ 10 + 60 b. 30 + 8 ☐ 37

Part 4

a. 27 + 9 = ______ b. 124 + 9 = ______ c. 38 + 9 = ______

Part 5 Work the estimation problems.

a. $\begin{array}{r} 62 \\ +\ 17 \\ \hline \end{array}$ ☐

b. $\begin{array}{r} 13 \\ +\ 41 \\ \hline \end{array}$ ☐

c. $\begin{array}{r} 89 \\ -\ 18 \\ \hline \end{array}$ ☐

d. $\begin{array}{r} 57 \\ -\ 22 \\ \hline \end{array}$ ☐

Part 6

a. $\begin{array}{r} 18 \\ -\ 9 \\ \hline \end{array}$ b. $\begin{array}{r} 7 \\ +\ 9 \\ \hline \end{array}$ c. $\begin{array}{r} 16 \\ -\ 9 \\ \hline \end{array}$ d. $\begin{array}{r} 4 \\ +\ 9 \\ \hline \end{array}$ e. $\begin{array}{r} 9 \\ -\ 5 \\ \hline \end{array}$ f. $\begin{array}{r} 9 \\ -\ 6 \\ \hline \end{array}$

Copyright © The McGraw-Hill Companies, Inc.

Lesson 

Name ____________________

Part 1

a. 17 − 9 =

b. 13 − 9 =

c. 18 − 9 =

d. 12 − 9 =

e. 15 − 9 =

f. 19 − 9 =

g. 14 − 9 =

h. 11 − 9 =

i. 9 − 9 =

Part 2

   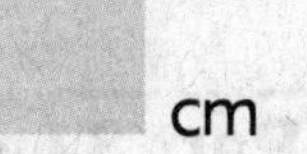 cm

 in.

Part 3

a. 408 − 122 =

b. 769 − 81 =

c. 517 − 293 =

Independent Work

Part 4

a. 152 + 56 =

b. 149 + 438 =

c. 762 + 195 =

d. 108 + 88 =

Copyright © The McGraw-Hill Companies, Inc.

Lesson 

Name ______________________

Part 5 Write the statement without the middle value.

a. P > R
16 > P

____

b. 24 < B
B < T

____

Part 6 Work each estimation problem.

a.
```
  37
+ 43
```

b.
```
  84
- 39
```

c.
```
  24
- 13
```

d.
```
  77
- 26
```

Part 7

a.
```
  17
-  9
```

b.
```
  15
-  9
```

c.
```
  12
-  9
```

d.
```
  14
-  5
```

e.
```
  14
-  4
```

Part 8

a. 9 + 30 + 108

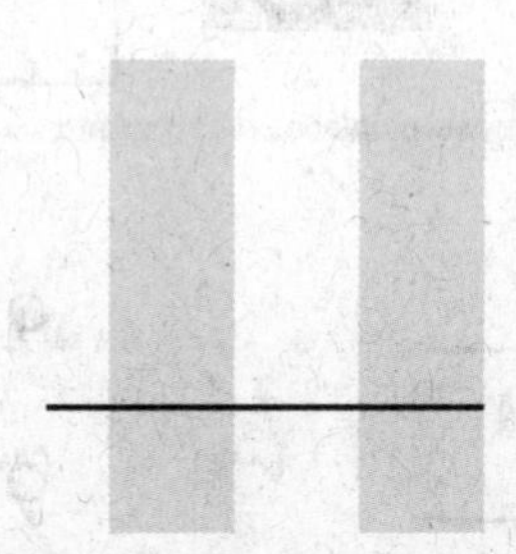

b. 23 + 400 + 7

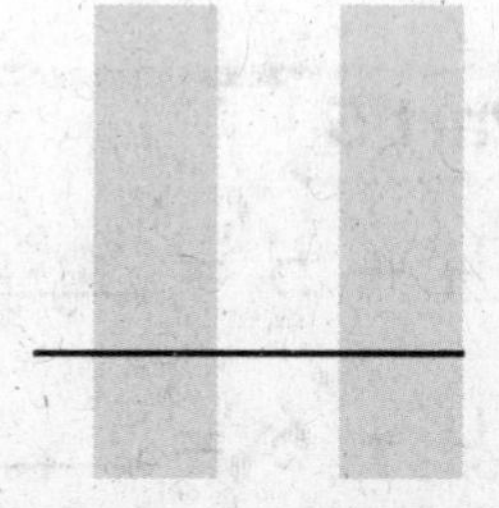

Copyright © The McGraw-Hill Companies, Inc.

Lesson 87

Name ______________________

Part 1

a. 440 − 123

b. 627 − 60

c. 915 − 24

d. 494 − 187

Part 2

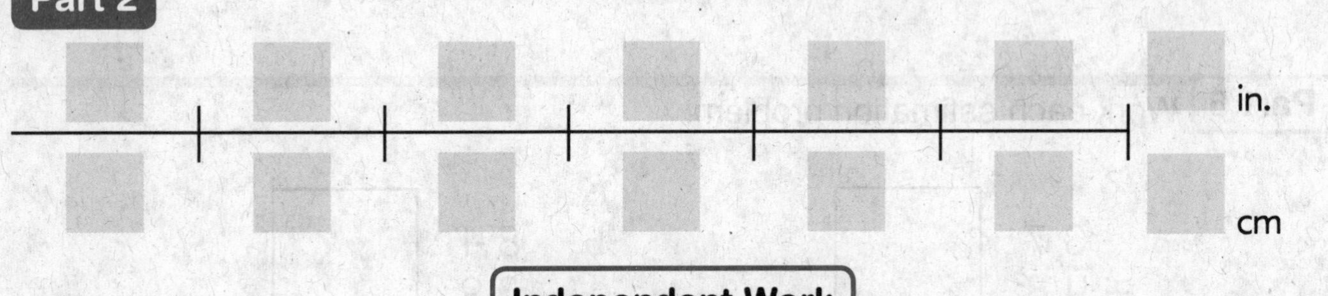

Independent Work

Part 3

Write the sign >, <, or =.

a. 40 + 1 ☐ 50 − 10

b. 10 + 52 ☐ 60 + 4

c. 30 + 30 ☐ 80 − 20

Part 4

Work each estimation problem.

a. 92 − 84

b. 68 − 57

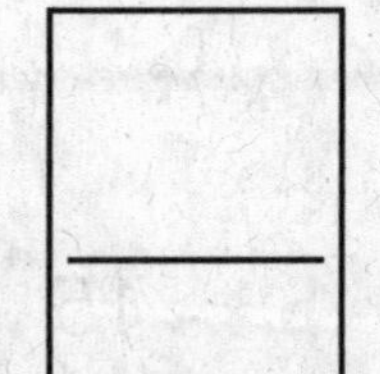

c. 43 + 28

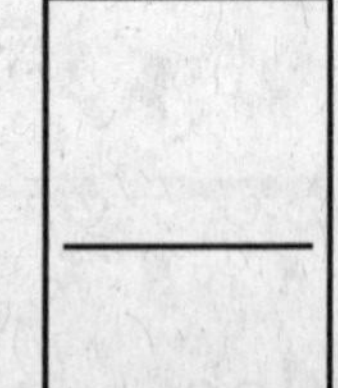

d. 19 + 71

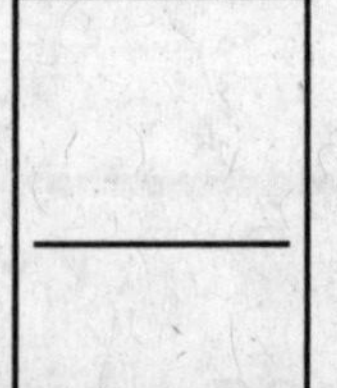

Part 5

a. 4 + 5 = ____

b. 19 − 9 = ____

c. 9 + 6 = ____

d. 4 + 6 = ____

e. 10 − 4 = ____

f. 9 − 4 = ____

Copyright © The McGraw-Hill Companies, Inc.

Lesson 

Name ______________________

Part 1

a. $\begin{array}{r} 605 \\ -183 \\ \hline \end{array}$

b. $\begin{array}{r} 451 \\ -\ \ 38 \\ \hline \end{array}$

c. $\begin{array}{r} 919 \\ -\ \ 28 \\ \hline \end{array}$

d. $\begin{array}{r} 650 \\ -635 \\ \hline \end{array}$

Part 2

a. 6 + 5 = ____	f. 15 − 10 = ____	k. 11 − 6 = ____	p. 14 − 5 = ____
b. 12 − 5 = ____	g. 10 − 5 = ____	l. 5 + 9 = ____	q. 12 − 7 = ____
c. 5 + 5 = ____	h. 5 + 7 = ____	m. 15 − 5 = ____	r. 5 + 10 = ____
d. 10 + 5 = ____	i. 5 + 6 = ____	n. 12 − 7 = ____	s. 11 − 5 = ____
e. 14 − 5 = ____	j. 9 + 5 = ____	o. 5 + 7 = ____	t. 6 + 5 = ____

Copyright © The McGraw-Hill Companies, Inc.

Independent Work

Part 3

a. 25 + 9 = ______ b. 32 + 9 = ______ c. 126 + 9 = ______

Part 4 Work each estimation problem.

a. $\begin{array}{r} 77 \\ +73 \\ \hline \end{array}$

b. $\begin{array}{r} 58 \\ -37 \\ \hline \end{array}$

Part 5

a. 7 + 4 = ____	b. 10 − 4 = ____	c. 6 + 9 = ____	d. 15 − 9 = ____
e. 11 − 7 = ____	f. 9 − 5 = ____	g. 10 − 5 = ____	h. 14 − 9 = ____

Lesson 

Name ____________________

Part 1

a. 15 − 10 = ____ f. 14 − 9 = ____ k. 6 + 5 = ____ p. 5 + 9 = ____

b. 5 + 7 = ____ g. 5 + 10 = ____ l. 12 − 5 = ____ q. 15 − 5 = ____

c. 10 − 5 = ____ h. 12 − 7 = ____ m. 5 + 5 = ____ r. 12 − 7 = ____

d. 9 + 5 = ____ i. 11 − 6 = ____ n. 14 − 5 = ____ s. 11 − 5 = ____

e. 5 + 6 = ____ j. 7 + 5 = ____ o. 10 + 5 = ____ t. 7 + 5 = ____

Part 2

a. $\begin{array}{r} 813 \\ -281 \\ \hline \end{array}$ b. $\begin{array}{r} 629 \\ +194 \\ \hline \end{array}$ c. $\begin{array}{r} 457 \\ +93 \\ \hline \end{array}$ d. $\begin{array}{r} 590 \\ -45 \\ \hline \end{array}$

Independent Work

Part 3

Write the sign >, <, or =.

a. 18 + 10 ☐ 40 − 20 b. 3 + 60 ☐ 72 − 10

c. 80 − 10 ☐ 60 + 10

Part 4

a. 10 − 4 = ____ b. 11 − 4 = ____ c. 9 − 7 = ____ d. 9 − 5 = ____

e. 10 − 6 = ____ f. 10 − 5 = ____ g. 16 − 9 = ____ h. 12 − 3 = ____

Copyright © The McGraw-Hill Companies, Inc.

Lesson 

Name ______________________

Part 1

a. 9 x ____ = 45   b. 4 x ____ = 28   c. 5 x ____ = 40

Independent Work

Part 2

Work each estimation problem.

a.
```
   4 6
 + 5 7
```

b.
```
   2 3
 + 5 7
```

Part 3

a. 5 + 7 = ____   c. 7 + 4 = ____   e. 5 + 6 = ____

b. 4 + 5 = ____   d. 6 + 4 = ____   f. 7 + 5 = ____

Part 4

Write the statement without the middle value.

a. $K < M$
$M < T$

____

b. $K < 18$
$T < K$

____

Part 5

a. 12 − 7   b. 6 + 5   c. 5 + 7   d. 11 − 6   e. 5 + 6   f. 12 − 5

g. 7 + 5   h. 11 − 5   i. 5 + 6   j. 12 − 7   k. 5 + 7   l. 11 − 6

Copyright © The McGraw-Hill Companies, Inc.

Lesson 91

Name ______________________

Part 1

a. 2 x ____ = 20 b. 4 x ____ = 20 c. 5 x ____ = 25 d. 10 x ____ = 60

Part 2

a. 26 hours ☐ 1 day

b. 10 minutes ☐ 1 hour

c. 12 inches ☐ 1 foot

d. 120 cents ☐ 1 dollar

e. 70 minutes ☐ 1 hour

f. 100 cents ☐ 1 dollar

Part 3

a. 9 – 6 = ____

b. 6 + 2 = ____

c. 6 + 5 = ____

d. 11 – 6 = ____

e. 10 – 6 = ____

f. 6 + 4 = ____

g. 6 + 6 = ____

h. 12 – 6 = ____

i. 6 + 4 = ____

j. 8 – 6 = ____

k. 6 + 5 = ____

l. 6 + 1 = ____

m. 10 – 6 = ____

n. 11 – 6 = ____

o. 6 + 3 = ____

Independent Work

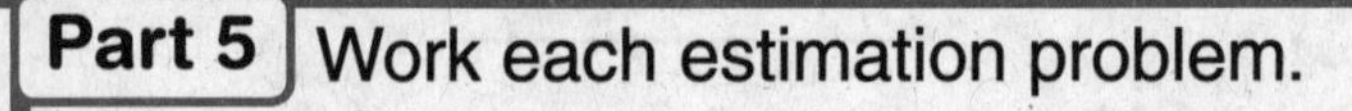

Part 4 Write the sign >, <, or =.

a. 20 + 30 ☐ 40 + 9

b. 80 – 10 ☐ 9 + 90

Part 5 Work each estimation problem.

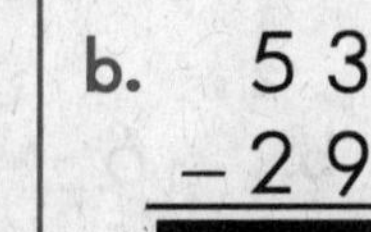

a. 37 + 61

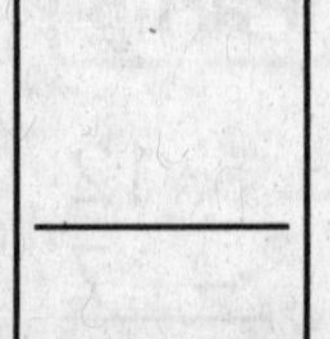

b. 53 – 29

Part 6

a. 9 – 5 = ____

b. 11 – 4 = ____

c. 10 – 6 = ____

d. 11 – 7 = ____

e. 12 – 5 = ____

Copyright © The McGraw-Hill Companies, Inc.

Lesson 

Name ______________________

Part 1

a. 5 x _____ = 40 b. 9 x _____ = 45 c. 10 x _____ = 40 d. 4 x _____ = 28

Part 2

a. $8.27 + 1.42 = ______

b. $6.89 − 2.18 = ______

c. $3.20 + 3.50 = ______

d. $4.87 − 3.35 = ______

Part 3

a. 13 inches ☐ 1 foot

b. 24 hours ☐ 1 day

c. 55 minutes ☐ 1 hour

d. 100 cents ☐ 1 dollar

e. 28 hours ☐ 1 day

f. 11 inches ☐ 1 foot

Part 4

a. 8 − 6 = _____ e. 11 − 6 = _____ i. 12 − 6 = _____ m. 11 − 5 = _____

b. 11 − 6 = _____ f. 7 − 6 = _____ j. 8 − 2 = _____ n. 10 − 6 = _____

c. 9 − 3 = _____ g. 11 − 5 = _____ k. 10 − 6 = _____ o. 7 − 6 = _____

d. 10 − 4 = _____ h. 10 − 4 = _____ l. 9 − 6 = _____

Copyright © The McGraw-Hill Companies, Inc.

Lesson 92

Name ____________________

Independent Work

Part 5 Write 4 facts for the family.

a. 7 ⟶ 4 ⟶ ■

____

____

____

____

Part 6 Work each estimation problem.

a. $\begin{array}{r} 83 \\ -26 \\ \hline \blacksquare \end{array}$ [___]

b. $\begin{array}{r} 48 \\ -31 \\ \hline \blacksquare \end{array}$ [___]

Part 7

a. 6 + 5 = ____ b. 7 + 5 = ____ c. 4 + 6 = ____

d. 4 + 5 = ____ e. 4 + 7 = ____

Part 8

a. $\begin{array}{r} 6 \\ +3 \\ \hline \end{array}$ b. $\begin{array}{r} 8 \\ -6 \\ \hline \end{array}$ c. $\begin{array}{r} 4 \\ +6 \\ \hline \end{array}$ d. $\begin{array}{r} 6 \\ +6 \\ \hline \end{array}$ e. $\begin{array}{r} 11 \\ -5 \\ \hline \end{array}$ f. $\begin{array}{r} 2 \\ +6 \\ \hline \end{array}$

g. $\begin{array}{r} 7 \\ -6 \\ \hline \end{array}$ h. $\begin{array}{r} 10 \\ -4 \\ \hline \end{array}$ i. $\begin{array}{r} 2 \\ +6 \\ \hline \end{array}$ j. $\begin{array}{r} 6 \\ -2 \\ \hline \end{array}$ k. $\begin{array}{r} 12 \\ -6 \\ \hline \end{array}$ l. $\begin{array}{r} 6 \\ +5 \\ \hline \end{array}$

Copyright © The McGraw-Hill Companies, Inc.

Lesson 

Name ____________________

Part 1

a. Linda had 44 stamps. She gave away 23 stamps. About how many stamps did she still have?

____

b. Frank had 12 stamps. Then his brother gave him 57 stamps. About how many stamps did Frank have altogether?

____

Part 2

a. 2 x ____ = 12 b. 5 x ____ = 20 c. 9 x ____ = 36 d. 4 x ____ = 32

Part 3

a. $\begin{array}{r} \$9.58 \\ -6.20 \\ \hline \end{array}$ b. $\begin{array}{r} \$1.07 \\ +8.02 \\ \hline \end{array}$ c. $\begin{array}{r} \$3.92 \\ +4.36 \\ \hline \end{array}$ d. $\begin{array}{r} \$8.90 \\ -5.42 \\ \hline \end{array}$

Part 4

a. 8 days ☐ 1 week

b. 10 months ☐ 1 year

c. 12 inches ☐ 1 foot

d. 110 cents ☐ 1 dollar

e. 60 minutes ☐ 1 hour

f. 23 hours ☐ 1 day

Independent Work

Part 5

Write the sign >, <, or =.

a. 34 + 10 ☐ 60 – 20

b. 9 + 60 ☐ 90 – 20

Part 6

a. 12 – 7 = ____ b. 11 – 6 = ____ c. 10 – 4 = ____

d. 9 – 5 = ____ e. 11 – 7 = ____

Copyright © The McGraw-Hill Companies, Inc.

Lesson 94

Name ______________________

Part 1

a. Hillary had 77 cards. Then she lost 18 cards. About how many cards did she still have?

____

b. Bill got 22 dollars last week. Then Bill got 26 dollars this week. About how many dollars did he have altogether?

____

Part 2

a. 10 x _____ = 50 b. 4 x _____ = 12 c. 2 x _____ = 14 d. 5 x _____ = 30

Part 3

a. $9.82 − 8.90

b. $5.52 − .16

c. $4.95 + 2.25

d. $1.40 + .90

Part 4

a. 15 months ☐ 1 year

b. 7 days ☐ 1 week

c. 23 hours ☐ 1 day

d. 58 minutes ☐ 1 hour

e. 14 inches ☐ 1 foot

f. 100 cents ☐ 1 dollar

Independent Work

Part 5

a. T < 17
17 < M

____

b. 200 > B
B > K

____

Part 6 Work each estimation problem.

a. 89 + 19

b. 43 + 27

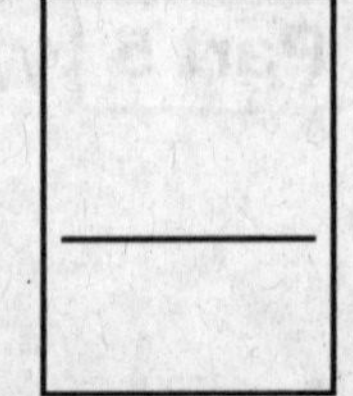

Part 7

a. 5 + 6 = _____

b. 7 + 5 = _____

c. 4 + 5 = _____

d. 12 − 5 = _____

e. 10 − 6 = _____

f. 9 − 4 = _____

g. 6 + 4 = _____

h. 7 + 4 = _____

i. 11 − 6 = _____

Copyright © The McGraw-Hill Companies, Inc.

Lesson 

Name ______________________

Part 1

a. 1 day ☐ 22 hours

b. 1 dollar ☐ 100 cents

c. 6 days ☐ 1 week

d. 13 months ☐ 1 year

e. 1 hour ☐ 60 minutes

f. 1 foot ☐ 13 inches

Part 2

a. $\begin{array}{r} \$7.52 \\ -3.19 \\ \hline \end{array}$

b. $\begin{array}{r} \$4.49 \\ +\ .15 \\ \hline \end{array}$

c. $\begin{array}{r} \$3.85 \\ -1.90 \\ \hline \end{array}$

d. $\begin{array}{r} \$5.08 \\ -4.25 \\ \hline \end{array}$

Independent Work

Part 3

a. $\begin{array}{r} 12 \\ -\ 5 \\ \hline \end{array}$

b. $\begin{array}{r} 6 \\ +4 \\ \hline \end{array}$

c. $\begin{array}{r} 9 \\ -5 \\ \hline \end{array}$

d. $\begin{array}{r} 11 \\ -\ 5 \\ \hline \end{array}$

e. $\begin{array}{r} 6 \\ +5 \\ \hline \end{array}$

f. $\begin{array}{r} 7 \\ +4 \\ \hline \end{array}$

g. $\begin{array}{r} 12 \\ -\ 7 \\ \hline \end{array}$

h. $\begin{array}{r} 6 \\ +5 \\ \hline \end{array}$

i. $\begin{array}{r} 4 \\ +5 \\ \hline \end{array}$

j. $\begin{array}{r} 5 \\ +3 \\ \hline \end{array}$

k. $\begin{array}{r} 10 \\ -\ 5 \\ \hline \end{array}$

l. $\begin{array}{r} 12 \\ -\ 5 \\ \hline \end{array}$

m. $\begin{array}{r} 4 \\ +7 \\ \hline \end{array}$

n. $\begin{array}{r} 9 \\ -5 \\ \hline \end{array}$

o. $\begin{array}{r} 5 \\ +5 \\ \hline \end{array}$

p. $\begin{array}{r} 10 \\ -\ 4 \\ \hline \end{array}$

q. $\begin{array}{r} 11 \\ -\ 7 \\ \hline \end{array}$

r. $\begin{array}{r} 7 \\ +5 \\ \hline \end{array}$

Copyright © The McGraw-Hill Companies, Inc.

Lesson 

Name ______________________________

Part 1

a. 1 yard ☐ 2 feet

b. 6 quarts ☐ 1 gallon

c. 55 minutes ☐ 1 hour

d. 1 week ☐ 7 days

e. 24 hours ☐ 1 day

f. 11 inches ☐ 1 foot

g. 1 dollar ☐ 85 cents

h. 1 year ☐ 14 months

Part 2

a. $\begin{array}{r} 9 \\ 5 \\ +\ 2 \\ \hline 17 \end{array}$ ☐

b. $\begin{array}{r} 10 \\ 4 \\ +\ 3 \\ \hline 17 \end{array}$ ☐

Independent Work

Part 3

a. $\begin{array}{r} 5 \\ +\ 6 \\ \hline \end{array}$ b. $\begin{array}{r} 7 \\ +\ 5 \\ \hline \end{array}$ c. $\begin{array}{r} 9 \\ +\ 4 \\ \hline \end{array}$ d. $\begin{array}{r} 6 \\ +\ 4 \\ \hline \end{array}$ e. $\begin{array}{r} 7 \\ +\ 6 \\ \hline \end{array}$ f. $\begin{array}{r} 5 \\ +\ 6 \\ \hline \end{array}$

g. $\begin{array}{r} 11 \\ -\ 6 \\ \hline \end{array}$ h. $\begin{array}{r} 9 \\ -\ 6 \\ \hline \end{array}$ i. $\begin{array}{r} 10 \\ -\ 6 \\ \hline \end{array}$ j. $\begin{array}{r} 8 \\ -\ 6 \\ \hline \end{array}$ k. $\begin{array}{r} 13 \\ -\ 6 \\ \hline \end{array}$ l. $\begin{array}{r} 5 \\ +\ 4 \\ \hline \end{array}$

Part 4

a. 56 + 9 = ____ b. 89 + 10 = ____ c. 76 + 9 = ____

Copyright © The McGraw-Hill Companies, Inc.

Lesson 

Name ____________________

Part 1

a. 1 gallon ☐ 3 quarts

b. 60 minutes ☐ 1 hour

c. 1 dollar ☐ 110 cents

d. 11 months ☐ 1 year

e. 1 foot ☐ 12 inches

f. 4 feet ☐ 1 yard

g. 1 day ☐ 25 hours

h. 8 days ☐ 1 week

Part 2

a. $\begin{array}{r} 2 \\ 4 \\ +\ 9 \\ \hline 16 \end{array}$ ☐

b. $\begin{array}{r} 4 \\ 1 \\ +\ 9 \\ \hline 13 \end{array}$ ☐

c. $\begin{array}{r} 8 \\ 4 \\ +\ 4 \\ \hline 16 \end{array}$ ☐

d. $\begin{array}{r} 9 \\ 3 \\ +\ 7 \\ \hline 19 \end{array}$ ☐

Part 3

☐

a. 7 + 6 = ____	f. 16 – 10 = ____	k. 13 – 7 = ____	p. 15 – 6 = ____
b. 13 – 6 = ____	g. 12 – 6 = ____	l. 6 + 9 = ____	q. 14 – 8 = ____
c. 6 + 6 = ____	h. 8 + 6 = ____	m. 16 – 6 = ____	r. 6 + 10 = ____
d. 10 + 6 = ____	i. 6 + 7 = ____	n. 14 – 8 = ____	s. 14 – 6 = ____
e. 15 – 9 = ____	j. 9 + 6 = ____	o. 6 + 8 = ____	t. 7 + 6 = ____

Copyright © The McGraw-Hill Companies, Inc.

Name ______________________

Independent Work

Part 4

a. $4 + 8$ = ___ b. $12 - 7$ = ___ c. $4 + 7$ = ___ d. $4 + 6$ = ___ e. $9 + 6$ = ___ f. $3 + 6$ = ___

g. $13 - 7$ = ___ h. $11 - 7$ = ___ i. $16 - 7$ = ___ j. $13 - 6$ = ___ k. $12 - 6$ = ___ l. $11 - 5$ = ___

Part 5

a. 141 + 10 = ______ b. 163 + 9 = ______ c. 148 + 10 = ______

Part 6

Work each estimation problem.

a. $77 + 23$

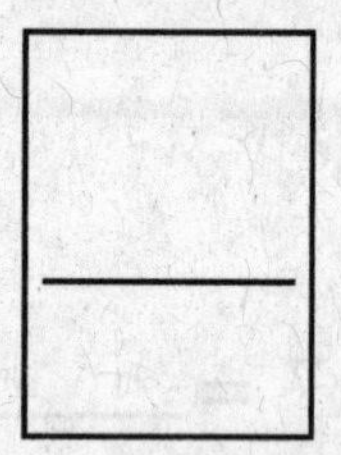

b. $49 + 62$

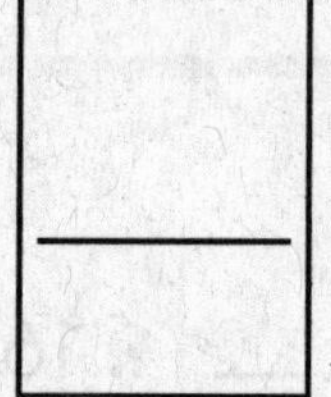

Copyright © The McGraw-Hill Companies, Inc.

Lesson 

Name ______________________________

Part 1

a. 1 gallon ☐ 5 quarts

b. 1 hour ☐ 60 minutes

c. 10 inches ☐ 1 foot

d. 1 foot ☐ 1 yard

e. 9 days ☐ 1 week

f. 1 day ☐ 24 hours

g. 14 months ☐ 1 year

h. 1 dollar ☐ 99 cents

Part 2

a. $8 + 1 + 9 = 18$ ☐

b. $8 + 2 + 6 = 15$ ☐

c. $2 + 1 + 8 = 12$ ☐

d. $10 + 5 + 5 = 20$ ☐

Part 3

☐

a. 16 – 10 = ____	f. 15 – 9 = ____	k. 7 + 6 = ____	p. 6 + 9 = ____
b. 6 + 8 = ____	g. 6 + 10 = ____	l. 14 – 6 = ____	q. 16 – 6 = ____
c. 12 – 6 = ____	h. 14 – 8 = ____	m. 6 + 6 = ____	r. 14 – 8 = ____
d. 9 + 6 = ____	i. 13 – 7 = ____	n. 15 – 6 = ____	s. 13 – 6 = ____
e. 6 + 7 = ____	j. 8 + 6 = ____	o. 10 + 6 = ____	t. 8 + 6 = ____

Copyright © The McGraw-Hill Companies, Inc.

Lesson 98

Name ____________________

Independent Work

Part 4

a. $\begin{array}{r} 10 \\ -\ 6 \\ \hline \end{array}$ b. $\begin{array}{r} 12 \\ -\ 6 \\ \hline \end{array}$ c. $\begin{array}{r} 11 \\ -\ 5 \\ \hline \end{array}$ d. $\begin{array}{r} 9 \\ -5 \\ \hline \end{array}$ e. $\begin{array}{r} 12 \\ -\ 7 \\ \hline \end{array}$ f. $\begin{array}{r} 10 \\ -\ 6 \\ \hline \end{array}$

g. $\begin{array}{r} 5 \\ +7 \\ \hline \end{array}$ h. $\begin{array}{r} 6 \\ +4 \\ \hline \end{array}$ i. $\begin{array}{r} 7 \\ +7 \\ \hline \end{array}$ j. $\begin{array}{r} 12 \\ -\ 5 \\ \hline \end{array}$ k. $\begin{array}{r} 4 \\ +7 \\ \hline \end{array}$ l. $\begin{array}{r} 9 \\ +4 \\ \hline \end{array}$

Part 5

Work each estimation problem.

a. $\begin{array}{r} 58 \\ +49 \\ \hline \end{array}$

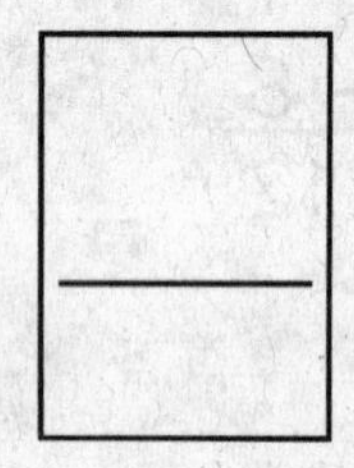

b. $\begin{array}{r} 26 \\ +32 \\ \hline \end{array}$

Copyright © The McGraw-Hill Companies, Inc.

Lesson  Name ____________________

Part 1

a. 1 week ☐ 6 days

b. 4 quarts ☐ 1 gallon

c. 1 year ☐ 18 months

d. 95 cents ☐ 1 dollar

e. 1 foot ☐ 12 inches

f. 1 yard ☐ 2 feet

g. 22 hours ☐ 1 day

h. 70 minutes ☐ 1 hour

Part 2

a. $\begin{array}{r} 8 \\ 7 \\ +\ 3 \\ \hline 18 \end{array}$ ☐

b. $\begin{array}{r} 8 \\ 8 \\ +\ 1 \\ \hline 16 \end{array}$ ☐

c. $\begin{array}{r} 3 \\ 1 \\ +\ 4 \\ \hline 9 \end{array}$ ☐

d. $\begin{array}{r} 5 \\ 3 \\ +\ 6 \\ \hline 14 \end{array}$ ☐

Independent Work

Part 3

a. 210 + 9 = ______ b. 10 + 138 = ______ c. 9 + 153 = ______

Copyright © The McGraw-Hill Companies, Inc.

Lesson 

Name ____________________

Part 1

a. 5 days ☐ 1 week

b. 1 dollar ☐ 100 cents

c. 1 gallon ☐ 3 quarts

d. 1 foot ☐ 10 inches

e. 3 feet ☐ 1 yard

f. 1 day ☐ 25 hours

g. 10 months ☐ 1 year

h. 1 hour ☐ 50 minutes

Part 2

a. 803 cents = ☐

b. 125 cents = ☐

c. 418 cents = ☐

Part 3

a. 11 − 7 = ____

b. 7 + 3 = ____

c. 7 + 6 = ____

d. 13 − 7 = ____

e. 9 − 7 = ____

f. 7 + 5 = ____

g. 7 + 7 = ____

h. 14 − 7 = ____

i. 7 + 5 = ____

j. 10 − 7 = ____

k. 7 + 6 = ____

l. 7 + 2 = ____

m. 12 − 7 = ____

n. 13 − 7 = ____

o. 7 + 4 = ____

Independent Work

Part 4

Check each answer. If an answer is wrong, cross it out and fix it.

a. $7 + 2 + 7 = 17$

b. $8 + 5 + 5 = 18$

c. $1 + 5 + 9 = 15$

d. $8 + 1 + 9 = 20$

Copyright © The McGraw-Hill Companies, Inc.

Lesson 

Name: ____________________

Part 5

a. $5 + 7$

b. $5 + 6$

c. $7 + 4$

d. $4 + 5$

e. $6 + 4$

f. $4 + 7$

g. $9 - 6$

h. $11 - 6$

i. $12 - 7$

j. $12 - 9$

k. $12 - 6$

l. $9 - 5$

Part 6

Write the statement without the middle value.

a. $97 > J$
$J > 80$

____

b. $P > B$
$B > R$

____

Part 7

Work each estimation problem.

a. $47 + 32$

b. $53 + 12$

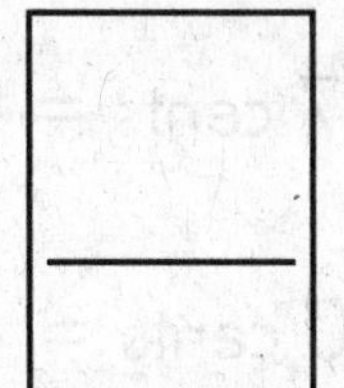

Part 8

a. $80 - 20 =$ _____

b. $70 - 10 =$ _____

c. $50 + 30 =$ _____

Copyright © The McGraw-Hill Companies, Inc.

Lesson 101

Name ____________________________

Part 1

	A	B	C
☆	12	17	10
△	5	6	7
☽	2	10	9

a. Circle the smallest number in the ☽ row.

b. Circle the largest number in the ☆ row.

c. Cross out the largest number in column B.

d. Cross out the smallest number in column A.

Part 2

a. 1 minute ☐ 50 seconds

b. 4 quarts ☐ 2 gallons

c. 60 seconds ☐ 1 minute

d. 1 meter ☐ 99 centimeters

e. 11 months ☐ 1 year

f. 2 feet ☐ 1 yard

g. 1 dollar ☐ 100 cents

h. 1 week ☐ 5 days

Part 3

a. 437 cents = ☐

b. 608 cents = ☐

c. 410 cents = ☐

d. 182 cents = ☐

Part 4

a. 10 – 7 = ____

b. 13 – 7 = ____

c. 11 – 4 = ____

d. 12 – 5 = ____

e. 13 – 7 = ____

f. 9 – 7 = ____

g. 13 – 6 = ____

h. 14 – 7 = ____

i. 12 – 5 = ____

j. 10 – 3 = ____

k. 12 – 7 = ____

l. 11 – 7 = ____

m. 13 – 6 = ____

n. 12 – 7 = ____

o. 9 – 2 = ____

Copyright © The McGraw-Hill Companies, Inc.

Lesson 101

Name ______________________

Independent Work

Part 5

a. 100 + 40 = ______ b. 70 + 20 = ______

Part 6

a.	b.	c.	d.	e.	f.
12 − 5	10 − 4	11 − 6	9 − 4	10 − 6	12 − 7

g.	h.	i.	j.	k.	l.
7 + 4	5 + 7	6 + 4	7 + 3	5 + 3	4 + 7

Part 7 Check each answer. If an answer is wrong, cross it out and fix it.

a. 2 + 6 + 3 = 11

b. 7 + 2 + 5 = 15

c. 5 + 5 + 2 = 12

Copyright © The McGraw-Hill Companies, Inc.

Lesson 

Name ______________________________

Part 1

	A	B	C
☆	8	5	11
△	2	10	6
☽	9	0	7

a. Circle the largest number in the △ row.

b. Circle the smallest number in the ☆ row.

c. Cross out the smallest number in column C.

d. Cross out the largest number in column A.

Part 2

a. 25 seconds ☐ 1 minute

b. 1 meter ☐ 65 centimeters

c. 1 week ☐ 9 days

d. 12 months ☐ 1 year

e. 110 centimeters ☐ 1 meter

f. 1 yard ☐ 3 feet

g. 1 gallon ☐ 3 quarts

h. 95 cents ☐ 1 dollar

Part 3

a. 908 cents = ☐

b. 340 cents = ☐

c. 779 cents = ☐

d. 215 cents = ☐

Independent Work

Part 4

Write the statement without the middle value.

a. $B > T$
$R > B$

____

b. $J < T$
$T < K$

____

Copyright © The McGraw-Hill Companies, Inc.

Lesson 102

Name ______________________

Part 5 Work each estimation problem.

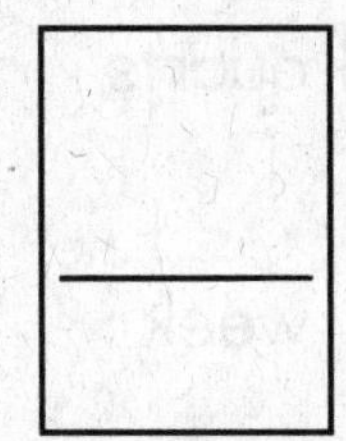

a.
```
   41
 + 77
```

b.
```
   66
 - 21
```

Part 6

a.	b.	c.	d.	e.	f.	g.
8 − 5	11 − 9	10 − 6	9 − 4	12 − 7	11 − 8	3 + 9

h.	i.	j.	k.	l.	m.
8 + 3	7 + 4	4 + 9	6 + 5	7 + 9	7 + 7

n.	o.	p.	q.	r.	s.
9 + 5	9 + 9	11 − 3	11 − 2	5 + 5	11 − 4

Copyright © The McGraw-Hill Companies, Inc.

Lesson 103

Name ____________________

Part 1

a. 1 meter ☐ 100 centimeters

b. 1 yard ☐ 5 feet

c. 70 seconds ☐ 1 minute

d. 1 year ☐ 14 months

e. 4 quarts ☐ 1 gallon

f. 1 week ☐ 6 days

g. 1 dollar ☐ 101 cents

h. 98 centimeters ☐ 1 meter

Independent Work

Part 2

Check each answer. If an answer is wrong, cross it out and fix it.

a. $\begin{array}{r} 1 \\ 9 \\ +\ 5 \\ \hline 16 \end{array}$ ☐

b. $\begin{array}{r} 2 \\ 4 \\ +\ 5 \\ \hline 11 \end{array}$ ☐

c. $\begin{array}{r} 3 \\ 3 \\ +\ 4 \\ \hline 12 \end{array}$ ☐

Part 3

a. $\begin{array}{r} 4 \\ +\ 4 \\ \hline \end{array}$
b. $\begin{array}{r} 11 \\ -\ 4 \\ \hline \end{array}$
c. $\begin{array}{r} 15 \\ -\ 6 \\ \hline \end{array}$
d. $\begin{array}{r} 20 \\ -\ 10 \\ \hline \end{array}$
e. $\begin{array}{r} 8 \\ +\ 8 \\ \hline \end{array}$
f. $\begin{array}{r} 8 \\ -\ 5 \\ \hline \end{array}$

g. $\begin{array}{r} 6 \\ +\ 4 \\ \hline \end{array}$
h. $\begin{array}{r} 6 \\ +\ 6 \\ \hline \end{array}$
i. $\begin{array}{r} 9 \\ -\ 3 \\ \hline \end{array}$
j. $\begin{array}{r} 10 \\ +\ 10 \\ \hline \end{array}$
k. $\begin{array}{r} 15 \\ -\ 6 \\ \hline \end{array}$
l. $\begin{array}{r} 18 \\ -\ 9 \\ \hline \end{array}$

Copyright © The McGraw-Hill Companies, Inc.

Lesson

Name ______________________________

Part 1

	A	B	C
☆			
△			
☽			

= ____ ____

= ____ ____

Part 2

a. 1 year ☐ 11 months

b. 75 seconds ☐ 1 minute

c. 1 day ☐ 28 hours

d. 7 days ☐ 1 week

e. 1 meter ☐ 110 centimeters

f. 11 feet ☐ 1 yard

g. 1 minute ☐ 60 seconds

h. 1 quart ☐ 1 gallon

Independent Work

Part 3

a. $B > R$
$R > T$

____

b. $J < T$
$5 < J$

____

Part 4

a. $40 - 30 =$ ______

b. $120 + 20 =$ ______

Part 5

a. $13 - 7$

b. $10 - 7$

c. $14 - 7$

d. $12 - 7$

e. $11 - 7$

f. $13 - 6$

g. $8 + 8$

h. $4 + 5$

i. $6 + 4$

j. $6 + 7$

k. $6 + 5$

l. $6 + 9$

Copyright © The McGraw-Hill Companies, Inc.

Lesson 

Name ______________________

Part 1

	A	B	C
☆			
△			
☽			

= ____ ____

= ____ ____

Part 2

a. 1 year ☐ 14 months

b. 25 hours ☐ 1 day

c. 3 feet ☐ 1 yard

d. 1 minute ☐ 90 seconds

e. 1 gallon ☐ 2 quarts

f. 95 centimeters ☐ 1 meter

g. 1 dollar ☐ 100 cents

h. 10 inches ☐ 1 foot

Independent Work

Part 3

a. 100 + 40 = ______

b. 80 + 30 = ______

Part 4

a. $9 - 5$

b. $11 - 4$

c. $12 - 5$

d. $10 - 5$

e. $11 - 5$

f. $14 - 5$

g. $5 + 7$

h. $6 + 6$

i. $4 + 7$

j. $6 + 5$

k. $4 + 5$

l. $5 + 6$

Part 5

a. 8 __ → 14

b. 7 __ → 12

c. 8 4 → __

d. 8 6 → __

e. 7 __ → 11

f. __ 4 → 10

Copyright © The McGraw-Hill Companies, Inc.

Lesson 

Name ______________________

Part 1

	pigs	goats	cows
farm A	20	1	50
farm B	13	32	21
farm C	0	9	18

a. 0 ____________ at farm ____________

b. 50 ____________ at farm ____________

c. 21 ____________ at farm ____________

Part 2

a.
$$\begin{array}{r} 66 \\ +46 \\ \hline \end{array}$$

____

____

b.
$$\begin{array}{r} 145 \\ +\ 35 \\ \hline \end{array}$$

____

____

Part 3

a. 8 + 4 = ____	f. 16 – 8 = ____	k. 12 – 8 = ____	p. 14 – 6 = ____
b. 13 – 5 = ____	g. 5 + 8 = ____	l. 6 + 8 = ____	q. 13 – 8 = ____
c. 8 + 3 = ____	h. 11 – 3 = ____	m. 16 – 8 = ____	r. 8 + 8 = ____
d. 8 + 8 = ____	i. 4 + 8 = ____	n. 13 – 8 = ____	s. 12 – 4 = ____
e. 14 – 6 = ____	j. 8 + 6 = ____	o. 5 + 8 = ____	t. 4 + 8 = ____

Copyright © The McGraw-Hill Companies, Inc.

Lesson 106

Name ______________________________

Independent Work

Part 4 Write the sign >, <, or =.

a. 60 seconds ☐ 1 minute

b. 5 quarts ☐ 1 gallon

c. 1 year ☐ 13 months

d. 1 week ☐ 9 days

e. 65 minutes ☐ 1 hour

f. 1 yard ☐ 3 feet

g. 120 cents ☐ 1 dollar

h. 1 meter ☐ 115 centimeters

Part 5 Write each answer.

a. $11 - 7 =$ ___

b. $9 - 6 =$ ___

c. $10 - 7 =$ ___

d. $11 - 5 =$ ___

e. $12 - 7 =$ ___

f. $9 - 4 =$ ___

g. $7 + 6 =$ ___

h. $9 + 7 =$ ___

i. $8 + 8 =$ ___

j. $4 + 7 =$ ___

k. $6 + 7 =$ ___

l. $5 + 7 =$ ___

Part 6 Write each missing number.

a. $8 \xrightarrow{6}$ ___

b. ___ $\xrightarrow{5} 11$

c. $7 \xrightarrow{\_\_} 13$

d. ___ $\xrightarrow{5} 12$

e. ___ $\xrightarrow{6} 13$

f. ___ $\xrightarrow{7} 14$

Copyright © The McGraw-Hill Companies, Inc.

Lesson 

Name ____________________

Part 1

	lions	tigers	bears
zoo X	12	6	5
zoo Y	3	10	2
zoo Z	8	4	0

a. 12 ______________ in zoo ______________

b. 4 ______________ in zoo ______________

c. 2 ______________ in zoo ______________

d. 10 ______________ in zoo ______________

Part 2

a. $\begin{array}{r} 174 \\ -117 \\ \hline \end{array}$

b. $\begin{array}{r} 99 \\ +104 \\ \hline \end{array}$

c. $\begin{array}{r} 256 \\ -\ \ 65 \\ \hline \end{array}$

______________ ______________ ______________

______________ ______________ ______________

Part 3

a. 16 – 8 = ____

b. 5 + 8 = ____

c. 11 – 3 = ____

d. 8 + 6 = ____

e. 4 + 8 = ____

f. 14 – 8 = ____

g. 8 + 8 = ____

h. 13 – 8 = ____

i. 14 – 6 = ____

j. 8 + 5 = ____

k. 8 + 4 = ____

l. 13 – 5 = ____

m. 8 + 3 = ____

n. 14 – 6 = ____

o. 8 + 8 = ____

p. 6 + 8 = ____

q. 16 – 8 = ____

r. 13 – 8 = ____

s. 12 – 4 = ____

t. 8 + 5 = ____

Part 4

a. ____ – 7 = 3 ______________

b. ____ – 2 = 9 ______________

c. ____ + 5 = 10 ______________

d. ____ – 5 = 10 ______________

e. ____ + 6 = 10 ______________

Copyright © The McGraw-Hill Companies, Inc.

Lesson 107

Name ______________________

Independent Work

Part 5 Write each missing number.

a. 6 —5→ ___

b. ___ —7→ 14

c. 8 —___→ 14

d. 6 —___→ 10

e. ___ —6→ 12

f. 5 —___→ 9

Part 6 Write the sign >, <, or =.

a. 1 foot ☐ 3 yards

b. 100 cents ☐ 2 dollars

c. 1 week ☐ 7 days

d. 12 months ☐ 1 year

e. 7 days ☐ 1 week

f. 1 minute ☐ 60 hours

g. 60 seconds ☐ 1 minute

h. 4 quarts ☐ 1 gallon

Part 7 Write each answer.

a. $8 + 6$

b. $6 + 7$

c. $7 + 4$

d. $7 + 7$

e. $8 + 8$

f. $6 + 6$

g. $9 + 5$

Part 8 Write each answer.

a. 170 + 20 = ______

b. 320 – 20 = ______

c. 130 + 60 = ______

d. 30 + 120 = ______

Copyright © The McGraw-Hill Companies, Inc.

Lesson  Name ____________________

Part 1

	dogs	cats	birds
barn	6	12	18
home	2	3	1
woods	4	5	80

a. 3 ____________ in the ____________

b. 18 ____________ in the ____________

c. 2 ____________ in the ____________

d. 80 ____________ in the ____________

Part 2

a. ____ − 3 = 9 ____________

b. ____ + 2 = 10 ____________

c. ____ − 7 = 9 ____________

d. ____ − 9 = 10 ____________

e. ____ + 6 = 12 ____________

Independent Work

Part 3 Write each missing number.

a. ___ 4 → 11

b. ___ 6 → 14

c. 7 ___ → 12

d. 7 ___ → 13

e. 8 ___ → 12

f. 9 ___ → 14

Part 4 Write each answer.

a. 14 − 8

b. 13 − 7

c. 13 − 4

d. 13 − 5

e. 14 − 6

f. 14 − 7

g. 12 − 7

h. 11 − 5

Part 5 Write each answer.

a. 20 + 150 = ______

b. 30 + 110 = ______

c. 50 − 40 = ______

d. 170 − 60 = ______

Copyright © The McGraw-Hill Companies, Inc.

Lesson 

Name ______________________

Part 1

a. 11 – 8 = ____	f. 8 + 4 = ____	k. 8 + 5 = ____
b. 8 + 2 = ____	g. 8 + 6 = ____	l. 8 + 1 = ____
c. 8 + 5 = ____	h. 14 – 6 = ____	m. 12 – 8 = ____
d. 13 – 8 = ____	i. 8 + 4 = ____	n. 13 – 8 = ____
e. 12 – 8 = ____	j. 10 – 8 = ____	o. 8 + 3 = ____

Independent Work

Part 2 Write each missing number.

a. __ 7 → 14

b. 8 6 → __

c. __ 8 → 16

d. __ 5 → 12

e. 7 6 → __

f. 9 __ → 15

Part 3 Write the sign >, <, or =.

a. 24 hours ☐ 1 minute

b. 1 year ☐ 12 months

c. 24 hours ☐ 1 day

d. 60 minutes ☐ 1 second

e. 1 week ☐ 12 days

f. 120 centimeters ☐ 1 meter

Part 4 Write each answer.

a. 13 – 7 = ____

b. 6 + 5 = ____

c. 9 + 5 = ____

d. 14 – 8 = ____

e. 6 + 8 = ____

f. 4 + 7 = ____

g. 6 + 9 = ____

h. 14 – 6 = ____

Copyright © The McGraw-Hill Companies, Inc.

Lesson 

Name ______________________

Part 1

a. 10 − 8 = ____	e. 13 − 8 = ____	i. 14 − 8 = ____	m. 13 − 5 = ____
b. 13 − 8 = ____	f. 9 − 8 = ____	j. 10 − 2 = ____	n. 12 − 8 = ____
c. 11 − 3 = ____	g. 13 − 5 = ____	k. 12 − 8 = ____	o. 14 − 6 = ____
d. 12 − 4 = ____	h. 12 − 4 = ____	l. 11 − 8 = ____	p. 9 − 1 = ____

Independent Work

Part 2 Write each answer.

a. 30 + 20 = ______

b. 150 − 30 = ______

c. 160 + 30 = ______

d. 170 − 20 = ______

Part 3 Complete each rule.

a. 3 feet = 1 ____________

b. 24 hours = 1 ____________

c. 12 months = 1 ____________

d. 1 gallon = ____ quarts

Part 4 Write each answer.

a. 146 + 10 = ______

b. 127 + 9 = ______

c. 148 + 9 = ______

d. 231 + 9 = ______

Copyright © The McGraw-Hill Companies, Inc.

Lesson 111

Name ______________________

Part 1

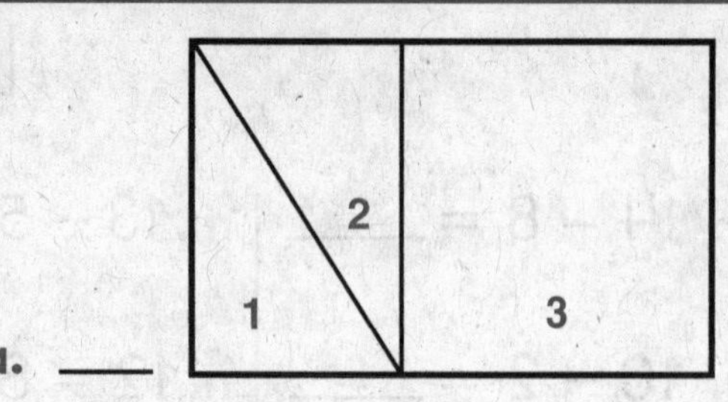

a. ___

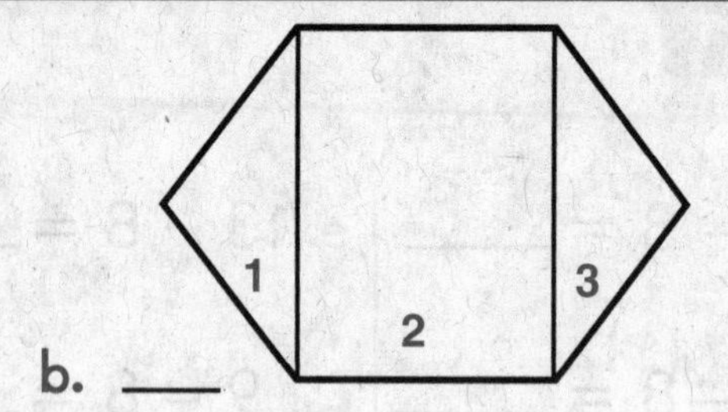

b. ___

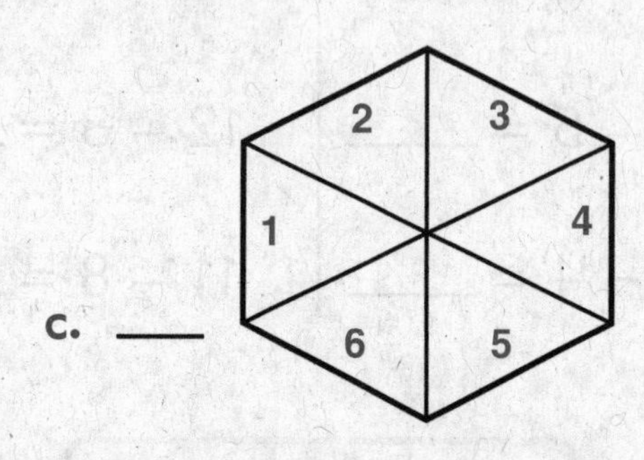

c. ___

Part 2

a. 15 – 8 = ____

b. 14 – 6 = ____

c. 8 + 7 = ____

d. 8 + 8 = ____

e. 15 – 7 = ____

f. 7 + 8 = ____

g. 8 + 6 = ____

h. 7 + 8 = ____

i. 6 + 8 = ____

j. 15 – 8 = ____

k. 8 + 7 = ____

l. 16 – 8 = ____

m. 15 – 8 = ____

n. 14 – 8 = ____

Independent Work

Part 3

a. 7 + 6

b. 5 + 8

c. 9 + 7

d. 8 + 4

e. 7 + 5

Part 4

a. 120 + 20 = ______

b. 120 – 20 = ______

c. 70 + 40 = ______

d. 70 – 40 = ______

Copyright © The McGraw-Hill Companies, Inc.

Lesson  Name ________

Part 1

a. 7 + 8 = ____	f. 15 − 8 = ____	k. 4 + 8 = ____	p. 12 − 8 = ____
b. 8 + 5 = ____	g. 8 + 6 = ____	l. 15 − 8 = ____	q. 14 − 8 = ____
c. 16 − 8 = ____	h. 13 − 8 = ____	m. 8 + 8 = ____	r. 12 − 4 = ____
d. 15 − 7 = ____	i. 8 + 7 = ____	n. 7 + 8 = ____	s. 5 + 8 = ____
e. 14 − 8 = ____	j. 8 + 4 = ____	o. 13 − 5 = ____	t. 14 − 6 = ____

Part 2

a. ____
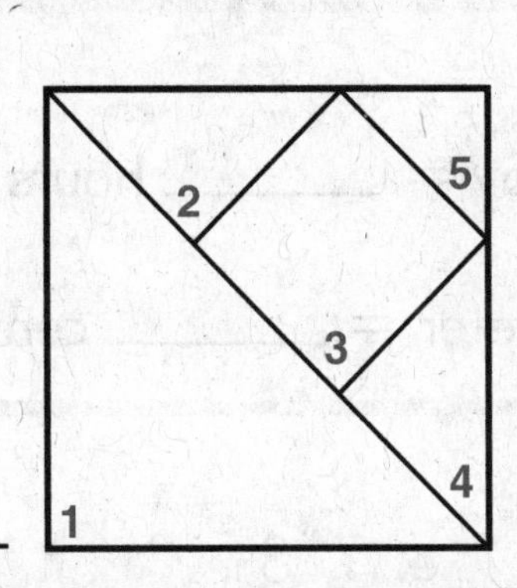

b. ____
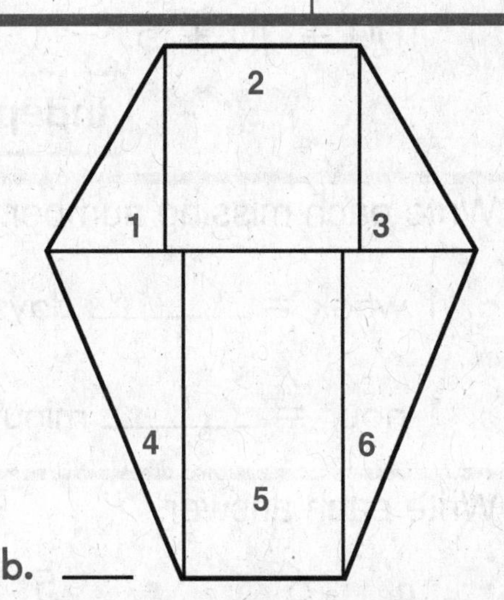

Part 3

a. ____

b. ____
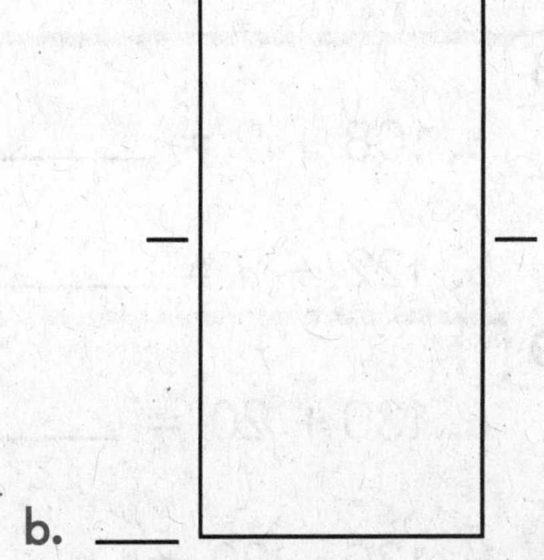

Copyright © The McGraw-Hill Companies, Inc.

Lesson 112

Name ______________________

Part 4

a. 301 ▢ 310 b. 518 ▢ 515 c. 625 ▢ 725

d. 620 ▢ 602 e. 110 ▢ 109 f. 123 ▢ 210

Part 5

a. 600 + 40 + 3 ▢ 648

b. 200 + 0 + 6 ▢ 260

c. 151 ▢ 100 + 10 + 5

d. 225 ▢ 300 + 0 + 5

e. 100 + 80 + 0 ▢ 179

Independent Work

Part 6 Write each missing number.

a. 1 week = _______ days

b. 1 hour = _______ minutes

c. 1 day = _______ hours

d. 1 meter = _______ centimeters

Part 7 Write each answer.

a. $8 + 9$ b. $6 + 8$ c. $5 + 8$ d. $7 + 6$ e. $3 + 8$ f. $5 + 6$

Part 8

a. 108 + 9 = _______

b. 122 + 9 = _______

c. 153 + 10 = _______

d. 441 + 9 = _______

Part 9

a. 130 + 20 = _______

b. 130 – 20 = _______

c. 400 + 50 = _______

d. 400 – 100 = _______

Copyright © The McGraw-Hill Companies, Inc.

Lesson 

Name ____________

Part 1

a. 7 + 8 = ___	f. 7 + 7 = ___	k. 17 − 10 = ___	p. 17 − 7 = ___
b. 9 + 7 = ___	g. 7 + 9 = ___	l. 14 − 7 = ___	q. 16 − 7 = ___
c. 10 + 7 = ___	h. 8 + 7 = ___	m. 15 − 7 = ___	r. 15 − 8 = ___
d. 8 + 7 = ___	i. 7 + 10 = ___	n. 16 − 9 = ___	s. 15 − 7 = ___
e. 7 + 9 = ___	j. 7 + 8 = ___	o. 15 − 8 = ___	t. 16 − 9 = ___

Part 2

Copyright © The McGraw-Hill Companies, Inc.

Part 3

a. 395 ☐ 401 b. 648 ☐ 651 c. 830 ☐ 803

d. 145 ☐ 205 e. 790 ☐ 709 f. 301 ☐ 299

Lesson 113

Name ______________________

Part 4

a. 148 ☐ 100 + 80 + 4

b. 600 + 70 + 8 ☐ 765

c. 710 ☐ 700 + 0 + 9

d. 515 ☐ 500 + 10 + 8

e. 300 + 20 + 0 ☐ 420

Independent Work

Part 5 Write each answer.

a. $\begin{array}{r} 8 \\ +\ 4 \\ \hline \end{array}$

b. $\begin{array}{r} 9 \\ +\ 6 \\ \hline \end{array}$

c. $\begin{array}{r} 4 \\ +\ 8 \\ \hline \end{array}$

d. $\begin{array}{r} 3 \\ +\ 8 \\ \hline \end{array}$

e. $\begin{array}{r} 8 \\ +\ 5 \\ \hline \end{array}$

f. $\begin{array}{r} 6 \\ +\ 7 \\ \hline \end{array}$

Part 6

a. 60 + 50 = ______

b. 60 – 50 = ______

c. 300 + 200 = ______

d. 300 – 200 = ______

Copyright © The McGraw-Hill Companies, Inc.

Lesson 114

Name ____________________

Part 1

a. 5 + 7 = ____	f. 8 + 6 = ____	k. 4 + 8 = ____	p. 6 + 5 = ____
b. 8 + 4 = ____	g. 4 + 6 = ____	l. 8 + 5 = ____	q. 6 + 8 = ____
c. 5 + 4 = ____	h. 5 + 8 = ____	m. 7 + 5 = ____	r. 7 + 4 = ____
d. 5 + 6 = ____	i. 4 + 7 = ____	n. 6 + 4 = ____	s. 8 + 7 = ____
e. 6 + 7 = ____	j. 7 + 8 = ____	o. 4 + 5 = ____	t. 7 + 6 = ____

Part 2

a. ___

b. ___

Independent Work

Part 3

Write each missing number.

a. 1 minute = ________ seconds

c. 1 foot = ________ inches

b. 1 gallon = ________ quarts

d. 1 meter = ________ centimeters

Part 4

a. 6 + 8

b. 5 + 9

c. 8 + 4

d. 7 + 6

e. 8 + 8

f. 5 + 8

Copyright © The McGraw-Hill Companies, Inc.

Lesson 

Name ______________________

Part 1

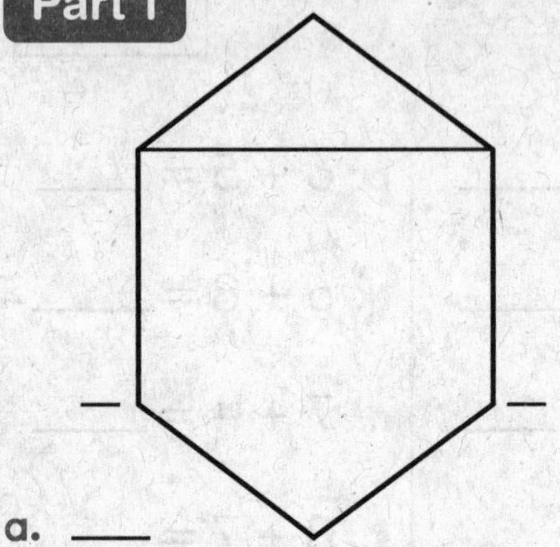

a. ___

b. ___

Part 2

a. 8 + 5 = ____	f. 11 − 4 = ____	k. 8 + 4 = ____	p. 7 + 4 = ____
b. 14 − 6 = ____	g. 4 + 6 = ____	l. 13 − 5 = ____	q. 12 − 5 = ____
c. 7 + 5 = ____	h. 15 − 7 = ____	m. 6 + 5 = ____	r. 6 + 8 = ____
d. 11 − 5 = ____	i. 12 − 4 = ____	n. 8 + 7 = ____	s. 10 − 4 = ____
e. 9 − 5 = ____	j. 7 + 6 = ____	o. 13 − 6 = ____	t. 5 + 4 = ____

Copyright © The McGraw-Hill Companies, Inc.

Part 3

cm

a. 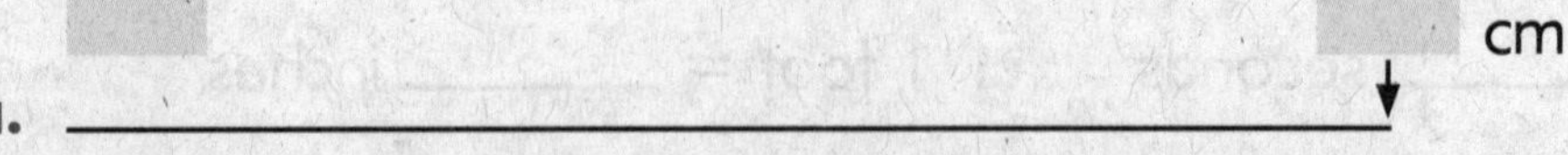

cm

b. 

big line: ______________________

difference: ______________________

Lesson 115

Name ______________________

Part 4

a. cm

b. cm

big line: ______________

difference: ______________

Independent Work

Part 5

Write each answer.

a. 9 + 8 = ____ c. 5 + 8 = ____ e. 3 + 8 = ____ g. 15 – 7 = ____

b. 15 – 6 = ____ d. 7 + 8 = ____ f. 14 – 8 = ____

Part 6

a. B < A
A < 36

____

b. R < 29
G < R

____

Part 7

Write the sign >, <, or =.

a. 100 ☐ 100 + 0 + 0 b. 199 ☐ 100 + 90 + 0

c. 200 + 80 + 9 ☐ 288 d. 501 ☐ 500 + 0 + 0

Copyright © The McGraw-Hill Companies, Inc.

Lesson 116

Name ____________________

Part 1

a.

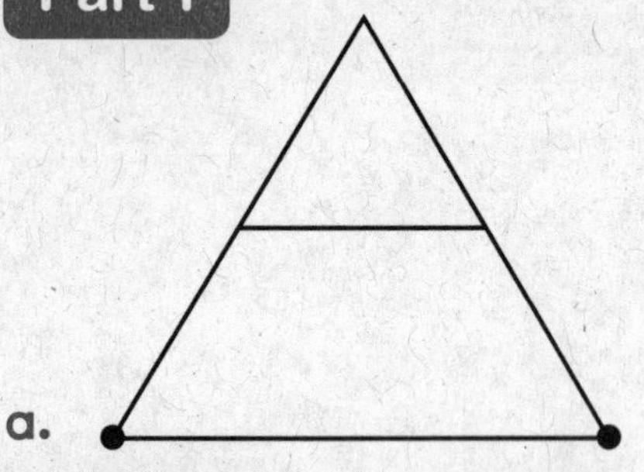

b.

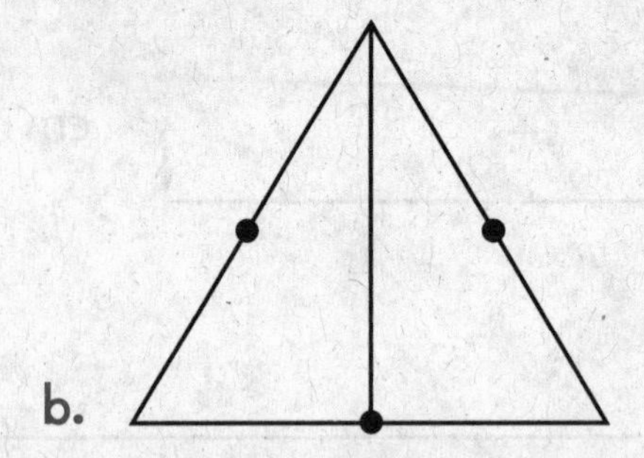

Part 2

a.

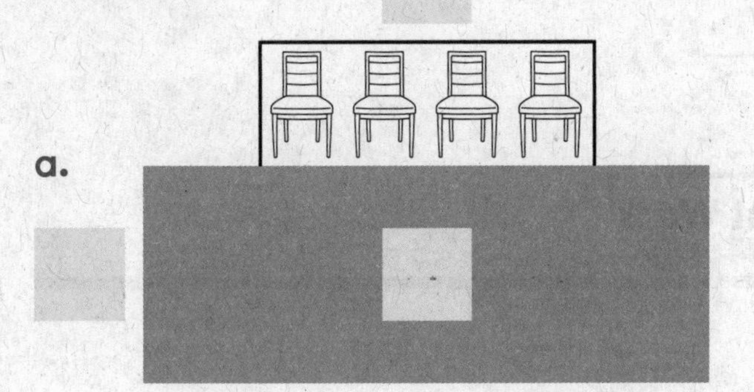

How many rows are there for 20 chairs?

____

b.

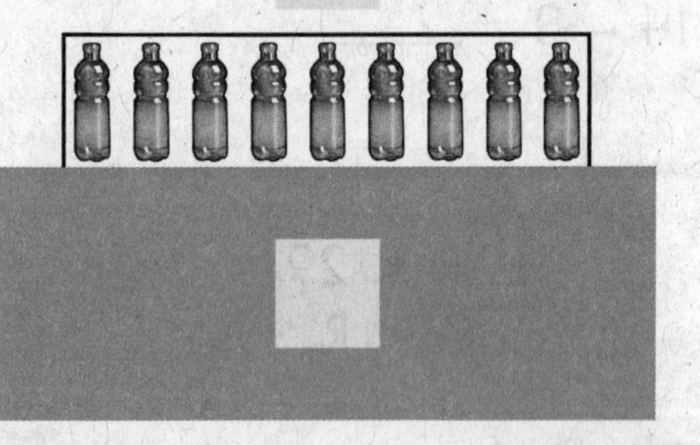

How many bottles are in 10 rows?

____

Part 3

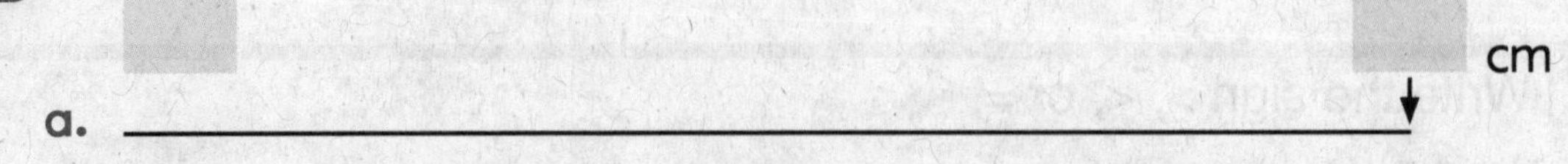

a. ________________________________ cm

b. ________________ cm

big line: ____________

difference: ____________

Copyright © The McGraw-Hill Companies, Inc.

Lesson 116

Name ____________________

Independent Work

Part 4 Write the sign >, <, or =.

a. 1 meter ☐ 104 centimeters

b. 1 quart ☐ 4 gallons

c. 1 day ☐ 24 hours

Part 5

a. 113 + 10 = ______

b. 106 + 9 = ______

c. 117 + 9 = ______

d. 148 + 9 = ______

Part 6 Write the statement without the middle value.

a. C > J
J > 12

______

b. 183 > T
C > 183

______

Part 7

a. 8 + 7 = ____	d. 8 + 4 = ____	g. 8 + 5 = ____	j. 7 + 5 = ____
b. 14 − 7 = ____	e. 15 − 8 = ____	h. 13 − 7 = ____	k. 16 − 8 = ____
c. 8 + 6 = ____	f. 12 − 5 = ____	i. 16 − 9 = ____	

Copyright © The McGraw-Hill Companies, Inc.

Lesson 117

Name ____________________

Part 1

a. 83 − 10 = ______

b. 41 − 10 = ______

c. 435 − 10 = ______

d. 280 − 10 = ______

e. 280 + 10 = ______

f. 280 + 100 = ______

Part 2

a. 

b. 

Part 3

basket A	🍎🍎🍎🍎🍎🍎🍎	
basket B	🍎🍎🍎🍎🍎🍎🍎🍎🍎🍎	
basket C	🍎🍎🍎🍎	
basket D	🍎🍎🍎🍎🍎	

Copyright © The McGraw-Hill Companies, Inc.

Lesson 117

Name ______________________________

Part 4

a.

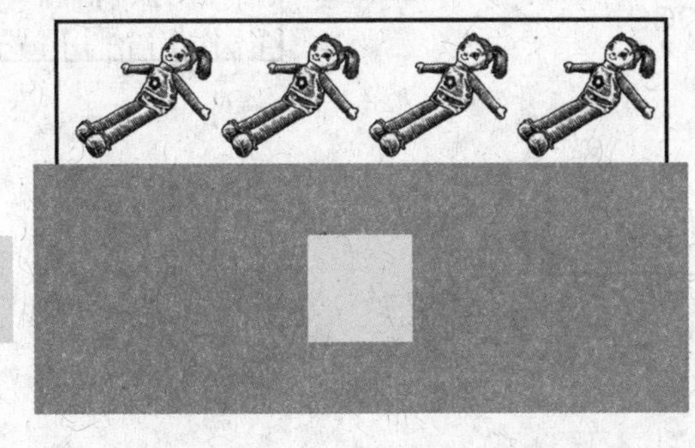

How many boxes are there for 40 toys?

____

b.

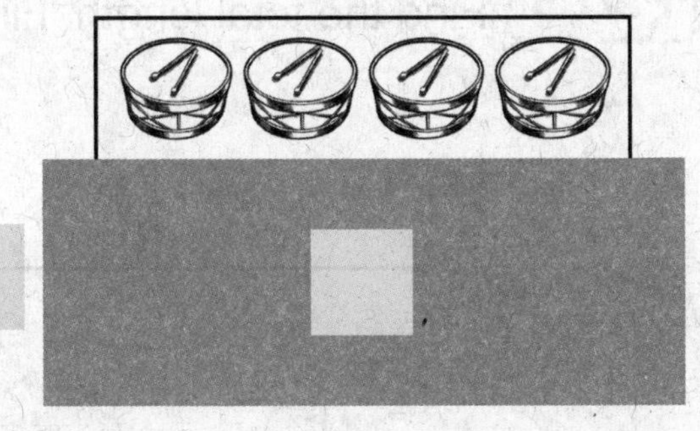

How many toys are in 10 boxes?

____

c.

How many boats are there for 30 children?

____

Copyright © The McGraw-Hill Companies, Inc.

Part 5

a. ⟶

b. ⟶

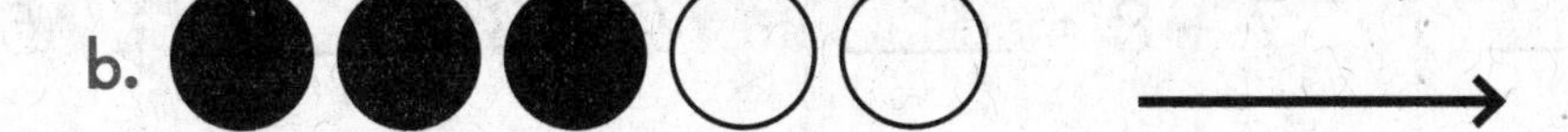

c. ⟶

d. ⟶

Lesson 117

Name ___________________________

Independent Work

Part 6 Find the total length. Find the difference.

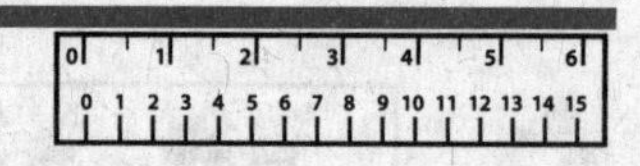

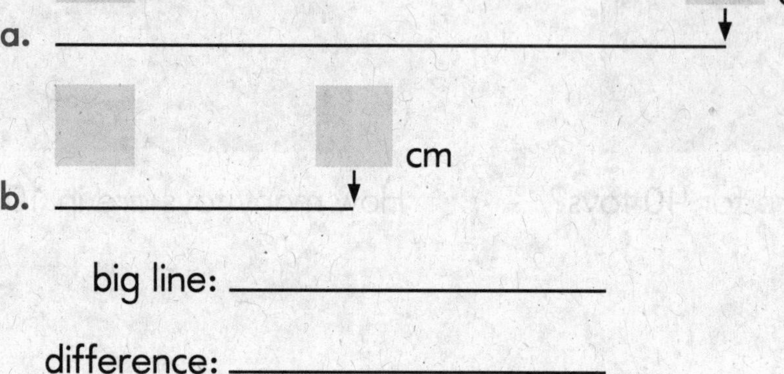

a. ______________ cm

b. ______________ cm

big line: ____________________

difference: ____________________

Part 7 Write the sign >, <, or =.

a. 8 days ☐ 1 week

b. 1 year ☐ 11 months

c. 1 hour ☐ 70 minutes

Part 8

a. $3 + 8 =$ ____	d. $7 + 8 =$ ____	g. $9 + 8 =$ ____	j. $5 + 8 =$ ____
b. $8 + 6 =$ ____	e. $15 - 8 =$ ____	h. $14 - 7 =$ ____	k. $17 - 8 =$ ____
c. $4 + 8 =$ ____	f. $12 - 8 =$ ____	i. $16 - 8 =$ ____	

Copyright © The McGraw-Hill Companies, Inc.

Lesson 

Name ____________________

Part 1

a. $680 + 10 =$ ______

b. $680 + 100 =$ ______

c. $680 - 10 =$ ______

d. $680 - 100 =$ ______

e. $251 + 100 =$ ______

f. $251 + 10 =$ ______

g. $251 - 10 =$ ______

h. $251 - 100 =$ ______

Part 2

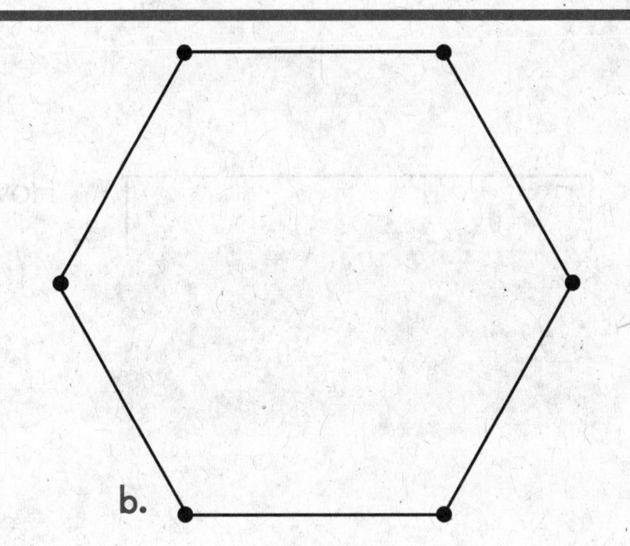

Part 3

shelf A		
shelf B		
shelf C		

Copyright © The McGraw-Hill Companies, Inc.

Lesson 118

Name ______________________

Part 4

a.

How many barns are there for 28 goats?

____

b.

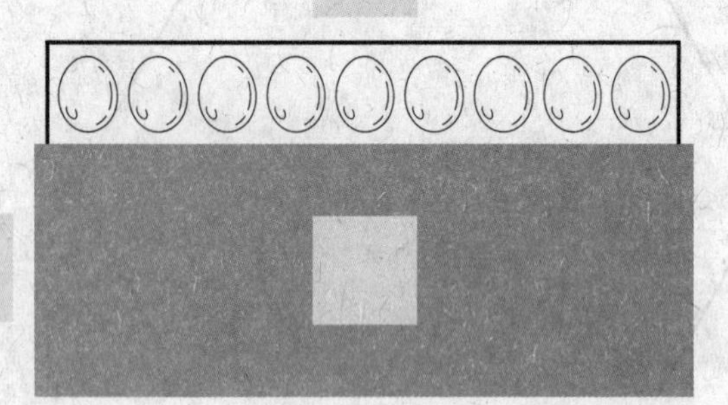

How many eggs are in 10 cartons?

____

c.

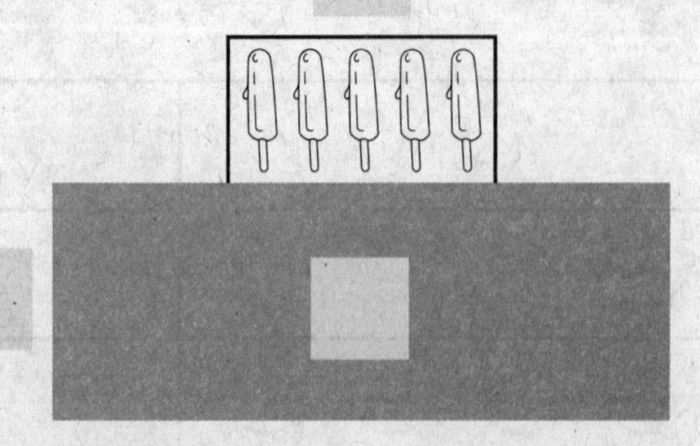

How many popsicles are in 12 boxes?

____

Copyright © The McGraw-Hill Companies, Inc.

Lesson 118

Name ____________________

Independent Work

Part 5 Find the total length. Find the difference.

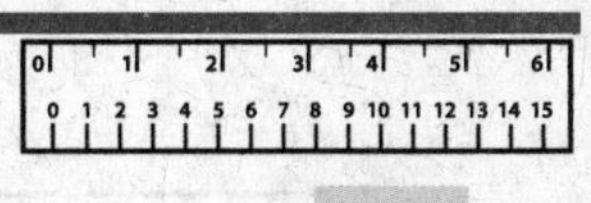

a. ____________________ cm

b. ____________ cm

big line: ____________

difference: ____________

Part 6 Write the sign >, <, or =.

a. 1 yard ☐ 2 feet b. 1 month ☐ 12 years c. 1 week ☐ 7 days

Part 7

a. $7 + 8 =$ ____ d. $16 - 7 =$ ____ g. $6 + 7 =$ ____

b. $7 + 7 =$ ____ e. $8 + 5 =$ ____ h. $15 - 7 =$ ____

c. $14 - 7 =$ ____ f. $15 - 9 =$ ____ i. $8 + 6 =$ ____

Part 8 Write the statement without the middle value.

a. $B > 86$
$86 > T$

____

b. $94 < T$
$R < 94$

____

Part 9

a. $124 + 9 =$ ______ b. $151 + 9 =$ ______ c. $136 + 10 =$ ______

Copyright © The McGraw-Hill Companies, Inc.

Lesson 119

Name ___________________________

Part 1

a.

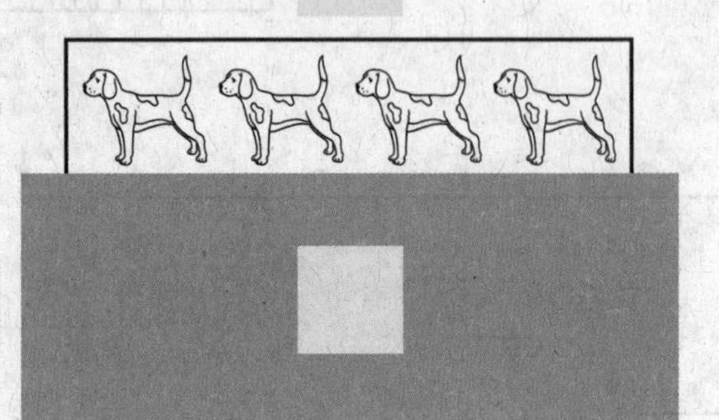

How many dogs are on 7 farms?

____

b.

How many boxes are there for 36 cups?

____

c.

How many trucks are there for 30 goats?

____

Copyright © The McGraw-Hill Companies, Inc.

Lesson 119

Name ______________________

Part 2

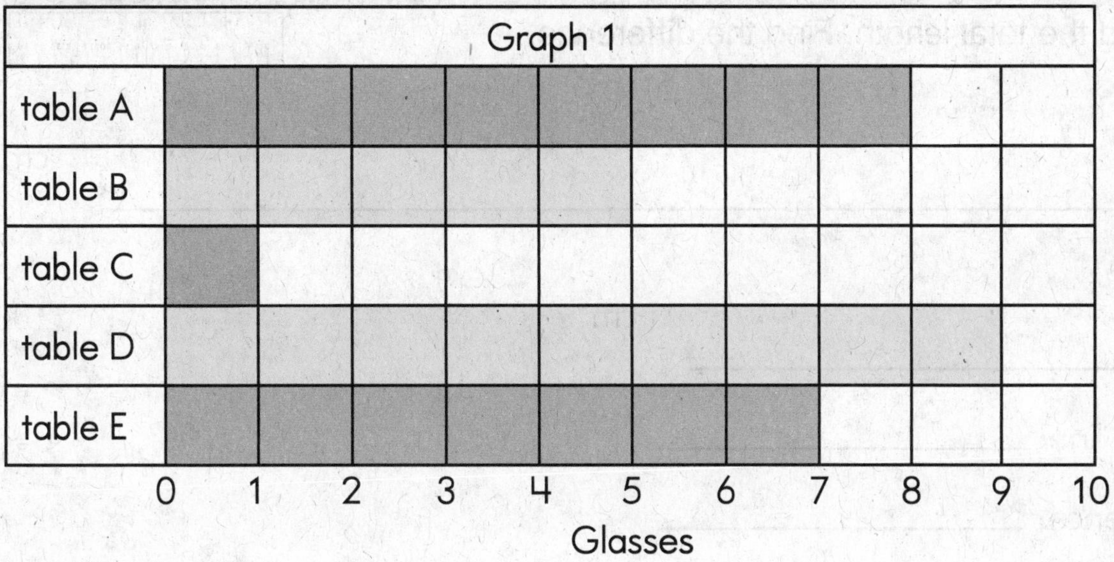

Graph 1

table A

table B

table C

table D

table E

0 1 2 3 4 5 6 7 8 9 10

Glasses

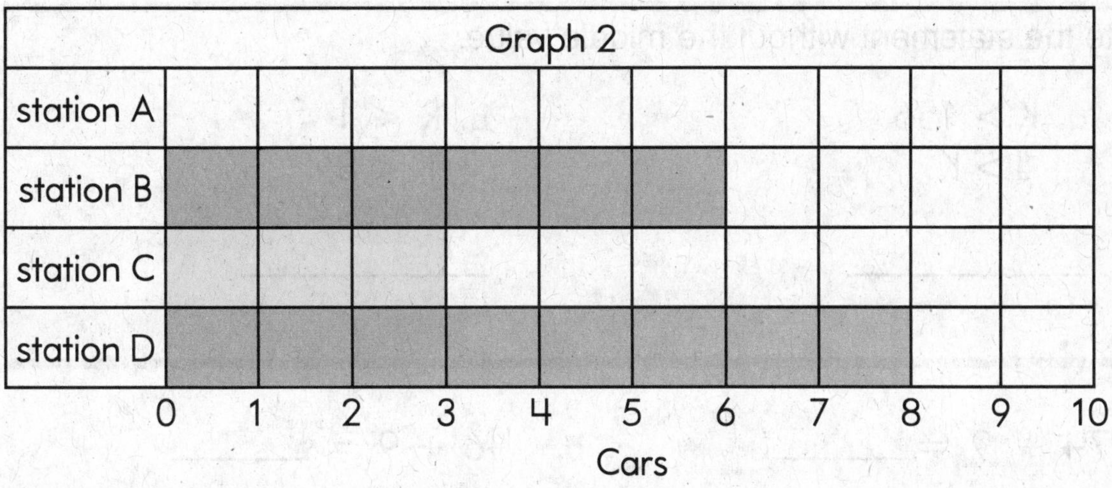

Graph 2

station A

station B

station C

station D

0 1 2 3 4 5 6 7 8 9 10

Cars

Copyright © The McGraw-Hill Companies, Inc.

Part 3

a. 236 + 100 = ______

b. 236 − 100 = ______

c. 236 − 10 = ______

d. 236 + 10 = ______

e. 850 − 10 = ______

f. 850 + 10 = ______

g. 850 − 100 = ______

h. 850 + 100 = ______

Lesson 119

Name ____________________________

Independent Work

Part 4 Find the total length. Find the difference.

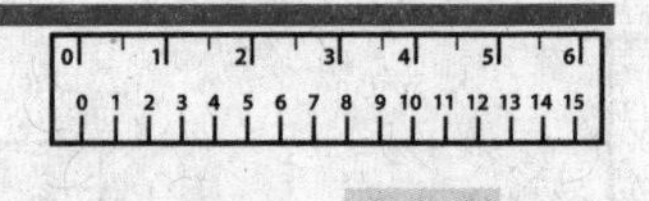

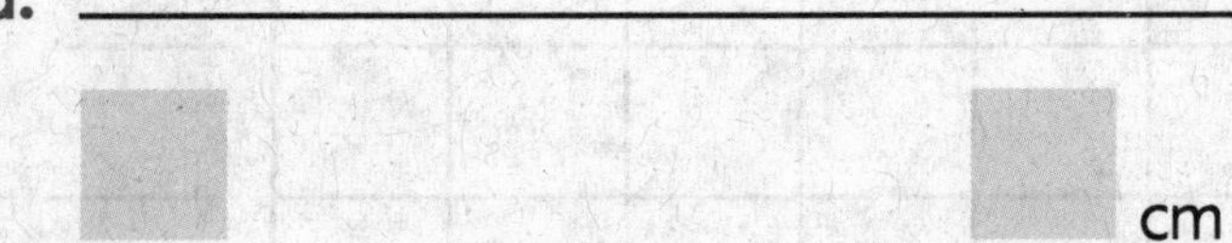

a. ______________________________ cm

b. ______________ cm

big line: ______________

difference: ______________

Part 5 Write the statement without the middle value.

a. $K > 156$
$J > K$

____

b. $R < 1$
$1 < B$

____

Part 6

a. $274 + 9 =$ ______

b. $46 + 9 =$ ______

c. $193 + 10 =$ ______

d. $197 + 9 =$ ______

Part 7 Write the sign >, <, or =.

a. 123 ☐ 100 + 30 + 2

b. 400 ☐ 400 + 0 + 0

c. 900 + 0 + 0 ☐ 899

d. 549 ☐ 500 + 40 + 8

Copyright © The McGraw-Hill Companies, Inc.

Lesson 

Name ______________________

Part 1

a. 412 − 100 = ______

b. 412 + 10 = ______

c. 412 + 100 = ______

d. 412 − 10 = ______

e. 738 + 100 = ______

f. 738 − 10 = ______

g. 738 + 10 = ______

h. 738 − 100 = ______

Part 2

closet A

closet B

closet C

closet D

0 1 2 3 4 5 6 7 8 9 10

Number of Coats

1. Which closet has the most coats? ______

2. Which closet has the fewest coats? ______

3. Which closets have more than 3 coats? ______

Independent Work

Part 3

Write the sign >, <, or =.

a. 1 minute ☐ 70 seconds

b. 1 hour ☐ 24 days

c. 1 yard ☐ 4 feet

d. 1 gallon ☐ 3 quarts

Copyright © The McGraw-Hill Companies, Inc.

Lesson 120

Name ____________________

Part 4 Measure each line. Answer the questions.

0 1 2 3 4 5 6
0 1 2 3 4 5 6 7 8 9 10 11 12 13 14 15

a. ______________________________ ☐ cm

b. ______________________________________ ☐ cm

How much longer is line b than line a? __________

How long are the two lines together? __________

Part 5

a. 7 + 9 = ____
b. 7 + 7 = ____
c. 6 + 8 = ____
d. 8 + 7 = ____

e. 3 + 7 = ____
f. 4 + 6 = ____
g. 6 + 7 = ____

h. 5 + 7 = ____
i. 7 + 9 = ____
j. 8 + 5 = ____

k. 5 + 9 = ____
l. 7 + 7 = ____
m. 6 + 5 = ____

Part 6 Write the sign >, <, or =.

a. 200 + 0 + 0 ☐ 210

b. 111 ☐ 100 + 10 + 0

c. 340 ☐ 300 + 0 + 4

d. 500 ☐ 500 + 0 + 0

e. 400 ☐ 395

f. 615 ☐ 651

g. 810 ☐ 811

Copyright © The McGraw-Hill Companies, Inc.

Lesson 121

Name ________________

Part 1

a. **Kay**

$58 – $25 ______

Bill

$58 – $27 ______

b. **Kay**

$65 – $20 ______

Bill

$65 – $15 ______

Part 2

a.

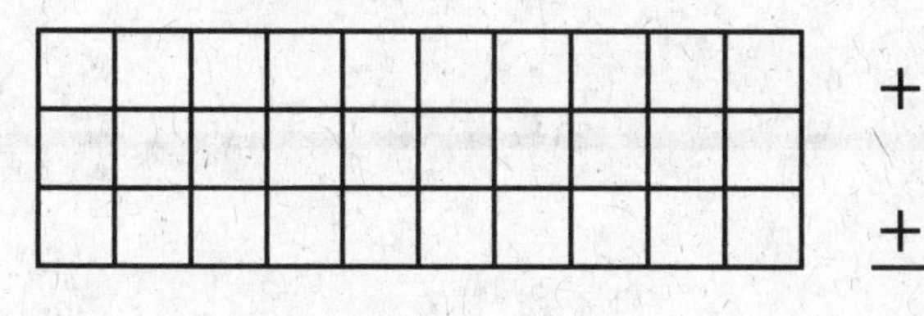

\+

\+

____

b.

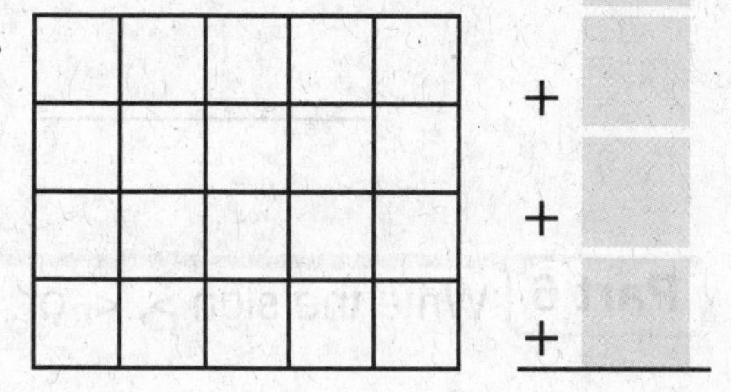

\+

\+

\+

____

Part 3

a. 175 + 100 = ______

b. 175 – 10 = ______

c. 175 – 100 = ______

d. 175 + 10 = ______

e. 249 – 100 = ______

f. 249 – 10 = ______

g. 249 + 100 = ______

h. 249 + 10 = ______

Copyright © The McGraw-Hill Companies, Inc.

Lesson 121

Name ______________________

Independent Work

Part 4 Write each answer.

a. $6 + 8 =$ ____

b. $16 - 8 =$ ____

c. $12 - 7 =$ ____

d. $9 + 8 =$ ____

e. $16 - 10 =$ ____

f. $14 - 7 =$ ____

g. $4 + 8 =$ ____

h. $11 - 5 =$ ____

i. $13 - 7 =$ ____

j. $7 + 8 =$ ____

k. $16 - 9 =$ ____

Part 5 Write each statement without the middle value.

a. $A < 46$
$46 < B$

____

b. $C > J$
$71 > C$

____

Part 6 Write the sign >, <, or =.

a. $800 + 0 + 0$ ☐ 801

b. 135 ☐ 153

c. 910 ☐ $900 + 0 + 1$

d. 299 ☐ $200 + 90 + 9$

Copyright © The McGraw-Hill Companies, Inc.

Lesson  Name ____________________

Part 1

a. **Kay** $36 – $25 ______

Bill $36 – $23 ______

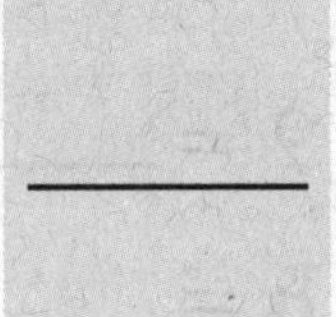

b. **Rob** $25 – $12 ______

Jan $25 – $21 ______

Part 2

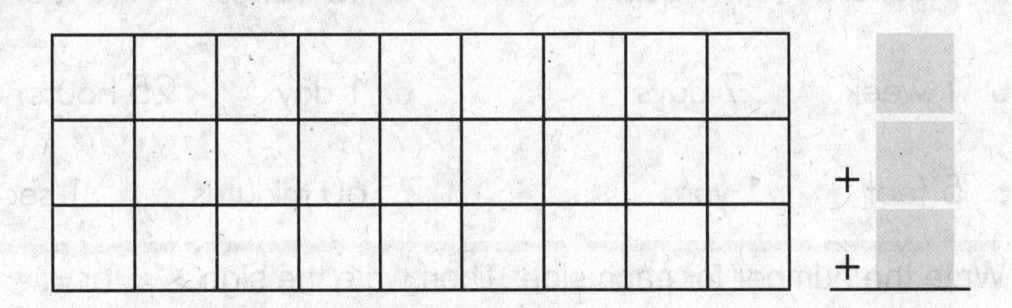

Part 3

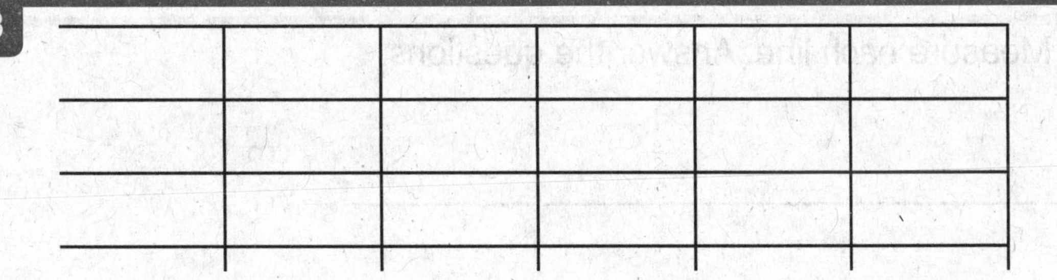

Number of Birds

- yard A has 4 birds
- yard B has 2 birds
- yard C has 5 birds

Copyright © The McGraw-Hill Companies, Inc.

Lesson 122

Name ______________________________

Independent Work

Part 4

a. 15 − 9 = ____	e. 7 + 9 = ____	i. 4 + 7 = ____	l. 7 + 5 = ____
b. 7 + 8 = ____	f. 6 + 7 = ____	j. 13 − 6 = ____	m. 14 − 8 = ____
c. 11 − 7 = ____	g. 12 − 5 = ____	k. 14 − 4 = ____	n. 15 − 8 = ____
d. 15 − 7 = ____	h. 15 − 10 = ____		

Part 5

Write the sign >, <, or =.

a. 4 gallons ☐ 1 quart

b. 1 week ☐ 7 days

c. 5 feet ☐ 1 yard

d. 13 inches ☐ 1 foot

e. 1 day ☐ 25 hours

f. 60 minutes ☐ 1 second

Part 6

Write the number for each side. Then write the sign >, <, or =.

a. 2 x 7 ☐ 9 + 6 b. 35 + 5 ☐ 5 x 7 c. 60 + 20 ☐ 10 x 7

Part 7

Measure each line. Answer the questions.

☐ cm

☐ cm

c. ____

How much longer is line c than line b? ____________

How much longer is line c than line a? ____________

Copyright © The McGraw-Hill Companies, Inc.

Lesson  Name ____________________

Part 1

a. Sam

$66 − $65 ______

Fran

$66 − $62 ______

Kay

$66 − $63 ______

b. Rob

$50 − $12 ______

Cal

$50 − $15 ______

Jan

$50 − $16 ______

Part 2

a.

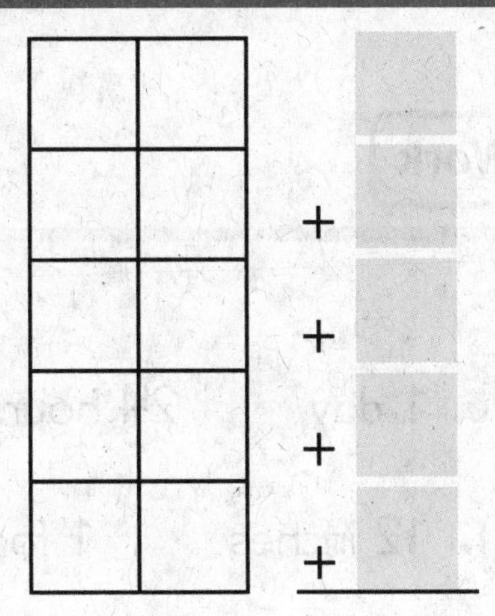

\+

\+

\+

\+

b.

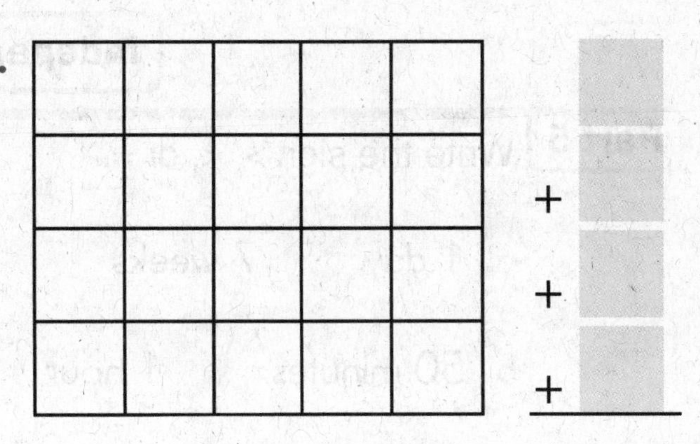

\+

\+

\+

____

____

Part 3

a. 58 − 2 = ______

b. 58 + 2 = ______

c. 3 + 63 = ______

d. 9 + 12 = ______

e. 75 − 5 = ______

f. 2 + 84 = ______

g. 99 − 1 = ______

h. 79 + 5 = ______

Copyright © The McGraw-Hill Companies, Inc.

Lesson 123

Name ______________________________

Part 4

___ ___ ___ ___ ___ ___ ___ ___ ___

- There are 6 pigs.
- There are 3 goats.
- There are 4 cows.

Independent Work

Part 5 Write the sign >, <, or =.

a. 1 day ☐ 7 weeks

b. 50 minutes ☐ 1 hour

c. 60 seconds ☐ 1 day

d. 1 gallon ☐ 3 quarts

e. 1 day ☐ 24 hours

f. 12 inches ☐ 1 foot

g. 7 days ☐ 1 week

Part 6 Write the statement without the middle value.

R < KB
KB < T

____

Part 7 Write the number for each side. Then write the sign >, <, or =.

a. 30 + 7 ☐ 33 + 4

b. 5 x 3 ☐ 8 + 8

Copyright © The McGraw-Hill Companies, Inc.

Lesson  Name ____________

Part 1

a.

May	Bob	Ray
\$99 − \$55 ______	\$99 − \$65 ______	\$99 − \$75 ______

b.

Rob	Cal	Jan
\$350 − \$190 ______	\$350 − \$150 ______	\$350 − \$180 ______

Part 2

\+

\+

\+

____

Part 3

a. ______ cm

d. ______ cm

b. ______ cm

e. ______ cm

c. ______ cm

f. ______ cm

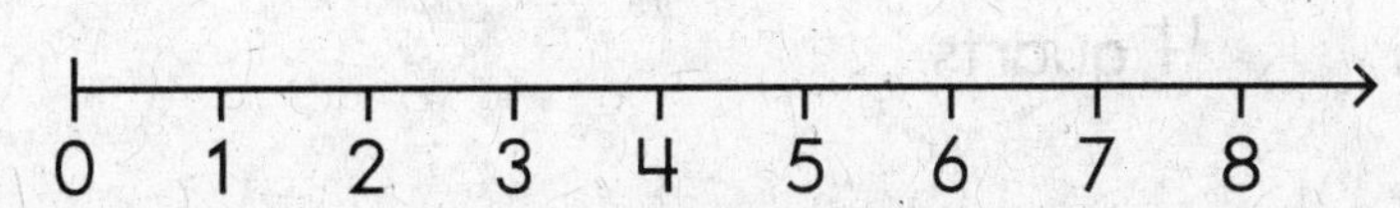

Copyright © The McGraw-Hill Companies, Inc.

Lesson 124

Name ______________________

Independent Work

Part 4 Write each answer.

a. 46 + 6 = ______

b. 9 + 27 = ______

c. 28 − 3 = ______

d. 4 + 13 = ______

e. 2 + 56 = ______

f. 88 − 4 = ______

g. 96 − 6 = ______

h. 95 + 5 = ______

Part 5 Write the statement without the middle value.

100 > B
K > 100

____

Part 6 Write the sign >, <, or =.

a. 40 + 8 ☐ 10 x 5

b. 2 x 9 ☐ 7 + 8

Part 7 Write the sign >, <, or =.

a. 100 centimeters ☐ 1 meter

b. 1 day ☐ 25 hours

c. 65 minutes ☐ 1 hour

d. 1 gallon ☐ 4 quarts

e. 3 yards ☐ 1 foot

f. 12 inches ☐ 1 foot

g. 1 minute ☐ 55 seconds

Copyright © The McGraw-Hill Companies, Inc.

Lesson 124

Name ______________________

Part 8 Measure each line. Answer the questions.

a. ________________________________ [ ] cm

b. ________________________________ [ ] cm

c. ____________________ [ ] cm

How much longer is line b than line a? ____________

How much longer is line a than line c? ____________

How long are lines b and c together? ____________

Part 9 Write each answer.

a. 43 + 9 = ______

b. 39 + 9 = ______

c. 46 + 8 = ______

d. 29 + 6 = ______

e. 57 + 7 = ______

f. 45 + 8 = ______

g. 57 + 8 = ______

h. 45 + 5 = ______

Copyright © The McGraw-Hill Companies, Inc.

Part 10 Write the sign >, <, or =.

a. 189 [ ] 100 + 80 + 9

b. 500 + 0 + 0 [ ] 510

c. 850 [ ] 800 + 40 + 5

d. 300 [ ] 300 + 0 + 0

Lesson 

Name ____________________

Part 1

a.

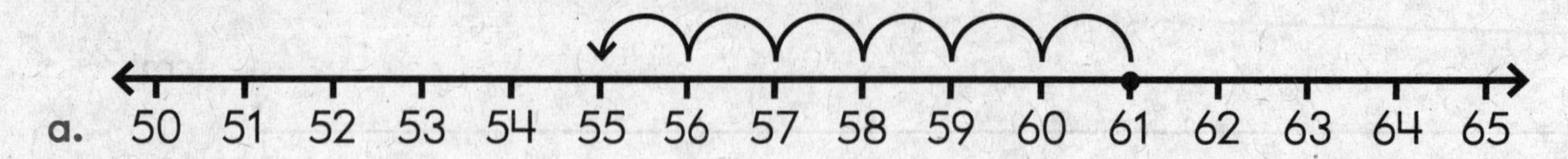

_______ _______ _______

Part 2

a. 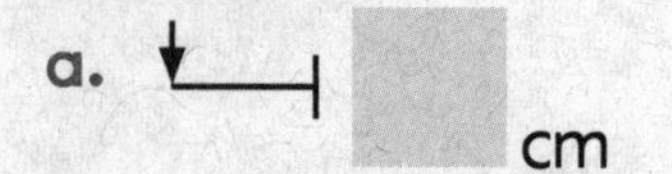 cm

e. cm

b. cm

f. cm

c. cm

g. cm

d.  cm

h. cm

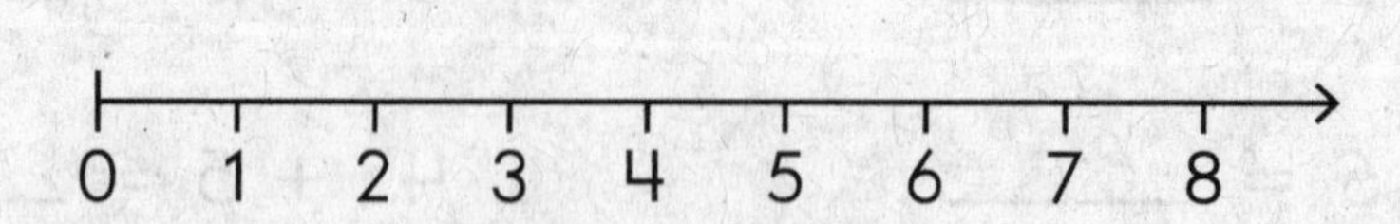

Copyright © The McGraw-Hill Companies, Inc.

 Name ____________

Part 3

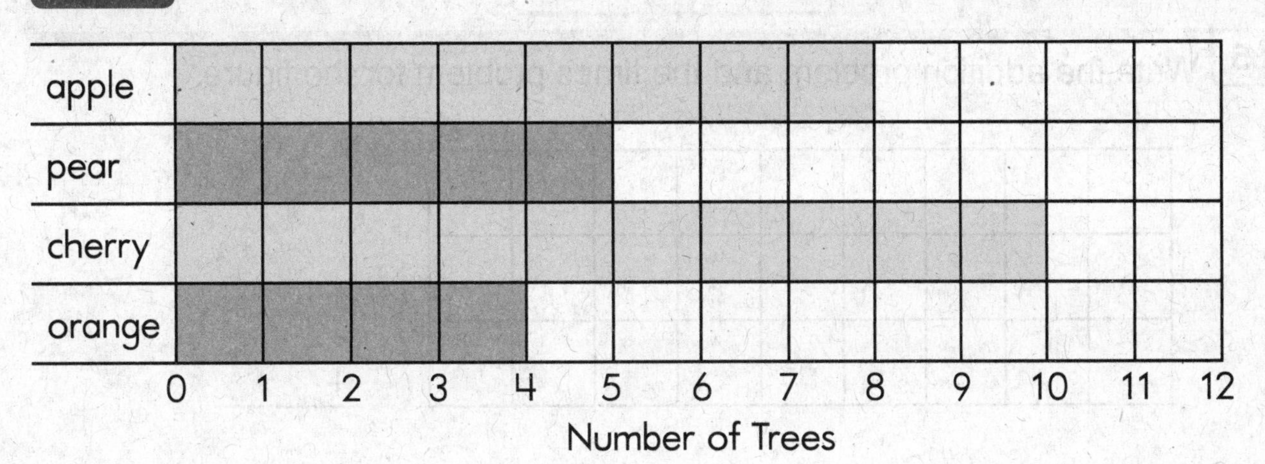

a. How many more cherry trees than apple trees are there? ______

b. How many fewer pear trees than cherry trees are there? ______

c. How many fewer orange trees than pear trees are there? ______

Part 4

a half a third a fourth a fifth

a.

How many parts are there? ______

What is each part called? ______

b.

How many parts are there? ______

What is each part called? ______

c.

How many parts are there? ______

What is each part called? ______

d.

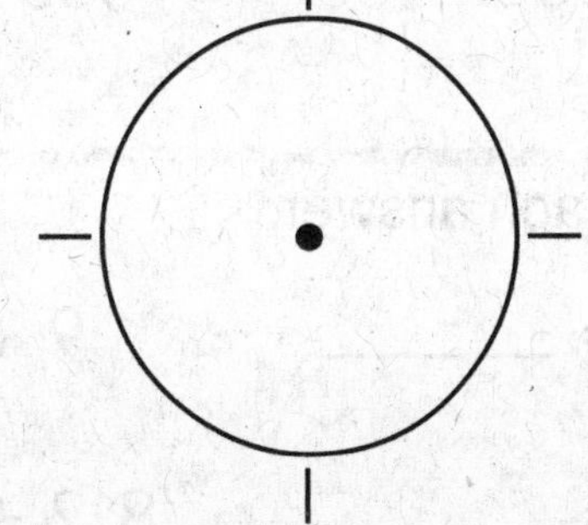

Copyright © The McGraw-Hill Companies, Inc.

Lesson 125

Name ______________________

Independent Work

Part 5 Write the addition problem and the times problem for the figure.

+

+

____

Part 6

a. 4 + 32 = ______ b. 5 + 45 = ______ c. 72 + 6 = ______

d. 99 − 7 = ______ e. 3 + 59 = ______ f. 66 − 4 = ______

Part 7 Write the sign >, <, or =.

a. 31 + 5 ☐ 38

b. 16 ☐ 28 − 10

c. 300 + 60 ☐ 360

d. 400 ☐ 600 − 300

e. 88 + 9 ☐ 98

Part 8 Write each answer.

a. 421 + 9 = ______ b. 9 + 34 = ______ c. 9 + 66 = ______

d. 9 + 184 = ______ e. 812 + 9 = ______

Copyright © The McGraw-Hill Companies, Inc.

Lesson 

Name ______________________________

Part 1

a.

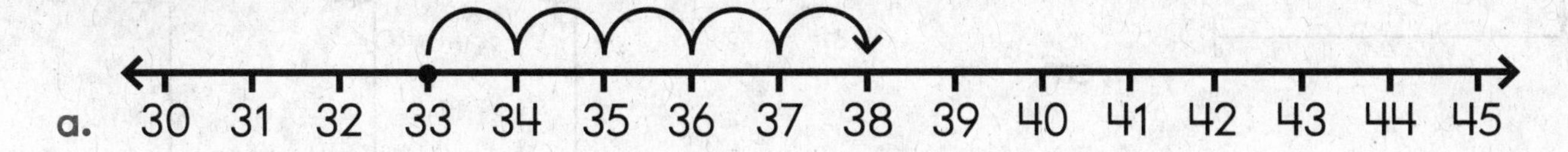

____

b.

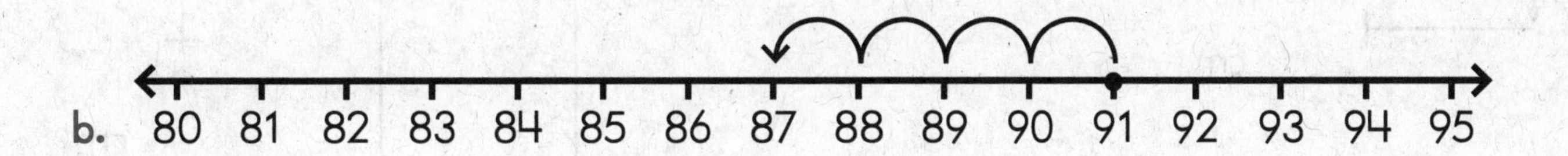

____

c.

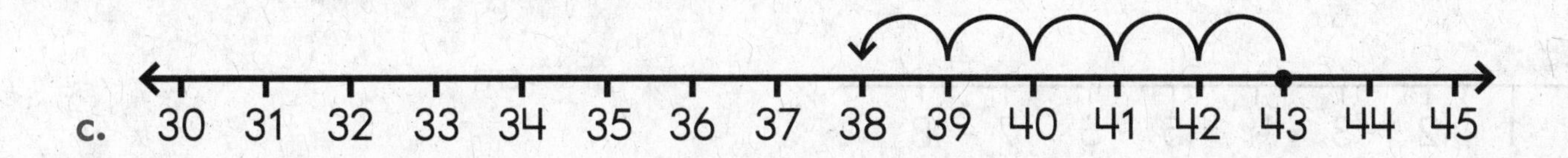

____

Part 2

a. 21 odd / even

b. 19 odd / even

c. 42 odd / even

d. 79 odd / even

e. 100 odd / even

f. 60 odd / even

g. 5 odd / even

Copyright © The McGraw-Hill Companies, Inc.

Lesson 126

Name ______________________

Part 3

a. ___ cm

b. ___ cm

c. ___ cm

d. ___ cm

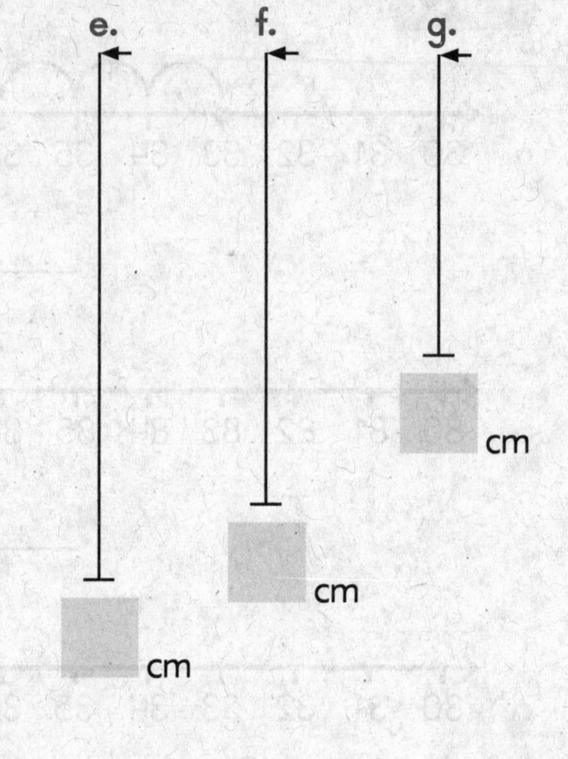

0 1 2 3 4 5 6 7 8

Part 4

Independent Work

Part 5

Write the sign >, <, or =.

a. 400 + 0 + 0 ☐ 401

b. 610 ☐ 600 + 0 + 1

c. 800 + 60 + 9 ☐ 871

d. 300 ☐ 300 + 0 + 0

Copyright © The McGraw-Hill Companies, Inc.

Lesson 126

Name ______________________________

Part 6 Answer each question.

a half	a third	a fourth	a fifth

a. ⊞

How many parts are there? ________

What is each part called? ___________

b. 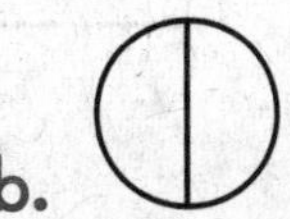

How many parts are there? ________

What is each part called? ___________

c. 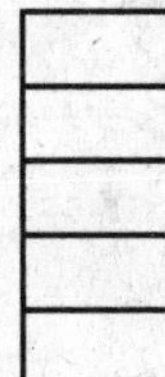

How many parts are there? ________

What is each part called? ___________

Part 7

a. 411 + 9 = ________ b. 79 + 9 = ________ c. 153 + 9 = ________

d. 9 + 87 = ________ e. 9 + 254 = ________

Part 8

a. 7 + 17 = ______ b. 5 + 29 = ______ c. 11 + 8 = ______

d. 64 − 4 = ______ e. 32 + 9 = ______ f. 68 − 5 = ______

Copyright © The McGraw-Hill Companies, Inc.

Lesson 

Name ____________________

Part 1

a. 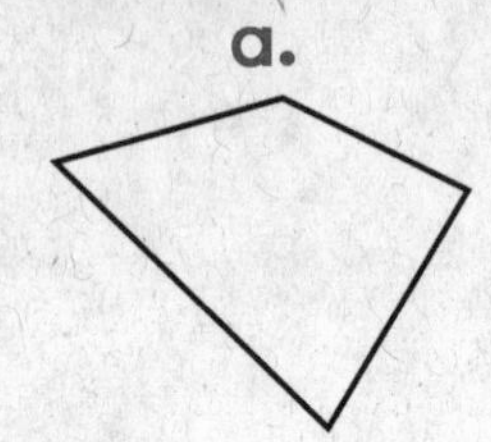

b.

c. 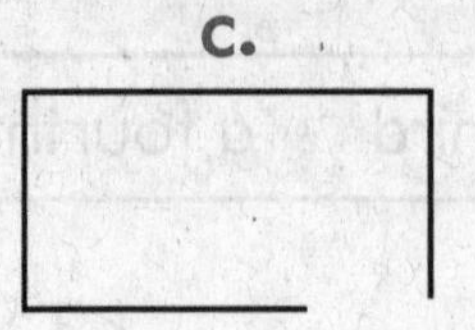

d.

e. 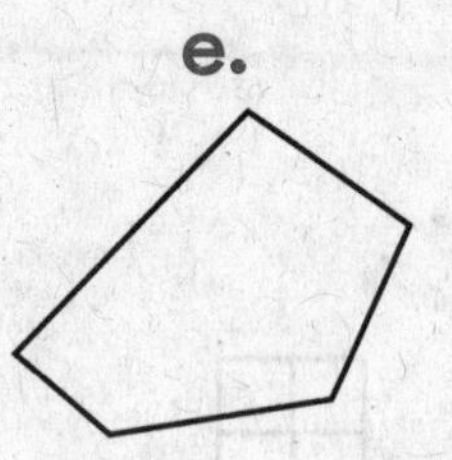

Part 2

a. 20 21 22 23 24 25 26 27 28 29 30 31 32 33 34 35

$27 + 6 =$ ______

b. 60 61 62 63 64 65 66 67 68 69 70 71 72 73 74 75

$68 + 4 =$ ______

c. 46 47 48 49 50 51 52 53 54 55 56 57 58 59 60

$53 - 6 =$ ______

Part 3

a. 59 odd / even

b. 50 odd / even

c. 8 odd / even

d. 38 odd / even

e. 17 odd / even

f. 21 odd / even

Copyright © The McGraw-Hill Companies, Inc.

Lesson 127

Name ____________________

Part 4

a. [ ] cm

b. [ ] cm

c. [ ] cm

d. [ ] cm

e. [ ] cm

f. [ ] cm

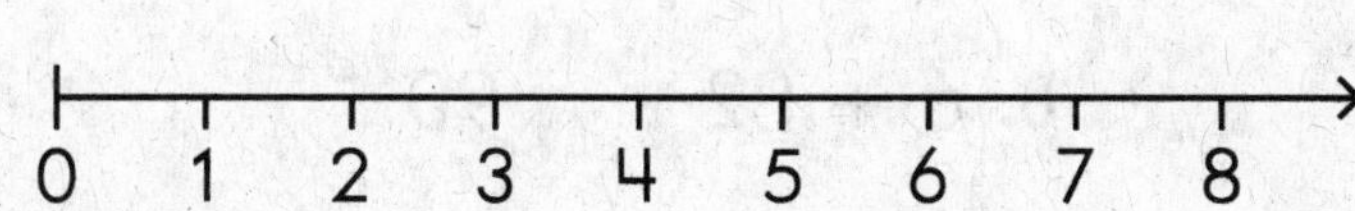

Part 5

a. 160 is ________ tens

b. 100 is ________ tens

c. 120 is ________ tens

d. 170 is ________ tens

Part 6

a.

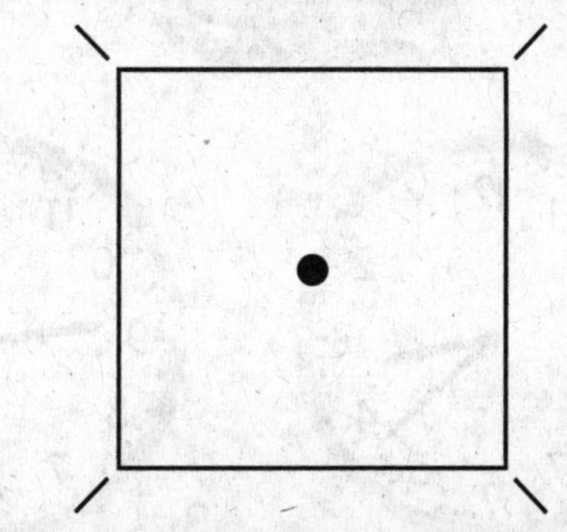

Copyright © The McGraw-Hill Companies, Inc.

Lesson 127

Name ______________________________

Independent Work

Part 7 Write the answer.

a. $48 + 6 =$ ____ b. $57 - 7 =$ ____ c. $59 - 4 =$ ____

d. $7 + 55 =$ ____ e. $8 + 38 =$ ____ f. $5 + 77 =$ ____

Part 8 Write the sign >, <, or =.

a. $600 + 12$ ☐ 611 d. 206 ☐ $300 - 100$

b. $6 + 82$ ☐ 90 e. $40 + 50$ ☐ 90

c. 45 ☐ $50 - 5$ f. $120 - 20$ ☐ 101 

Part 9 Write the number of the clock that shows the same time.

a. 

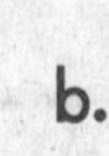
____

b. 

____

c. 

____

1. 

2. 

3. 

4. 

Copyright © The McGraw-Hill Companies, Inc.

Lesson 

Name ______________________

Part 1

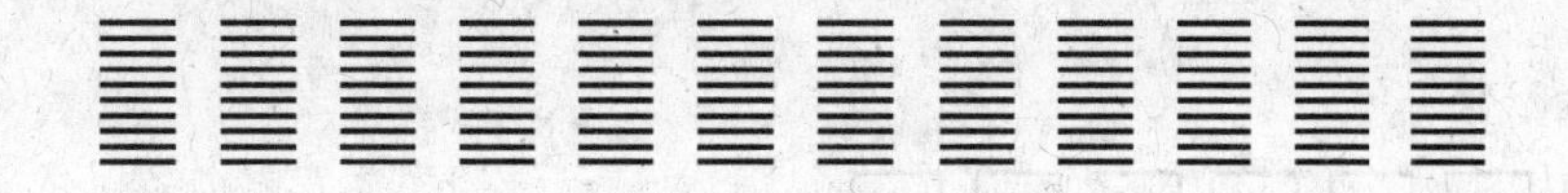

a. 100 + ☐ = ☐

b. ☐ + ☐ = ☐

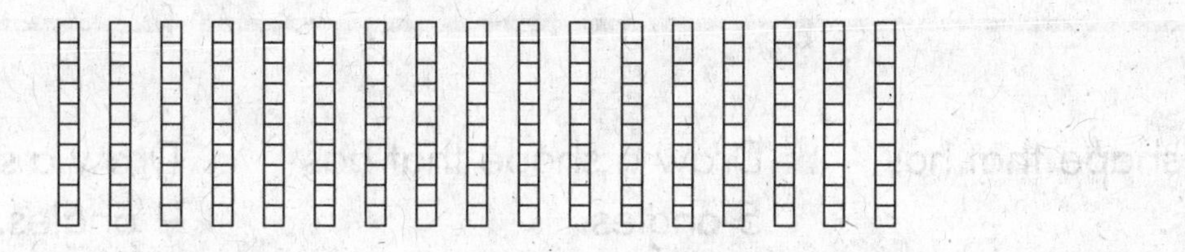

c. ☐ + ☐ = ☐

Part 2

a. 33 34 35 36 37 38 39 40 41 42 43 44 45 46 47 48 49 50

42 − 8 = ______

b. 12 13 14 15 16 17 18 19 20 21 22 23 24 25 26 27 28

19 + 5 = ______

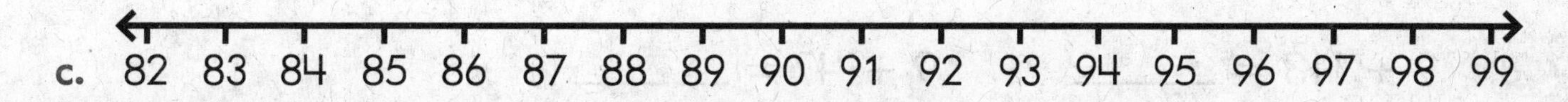

c. 82 83 84 85 86 87 88 89 90 91 92 93 94 95 96 97 98 99

90 − 7 = ______

Copyright © The McGraw-Hill Companies, Inc.

Lesson 128

Name ____________________

Part 3

a. □□|□□□□□□□□ odd / even ______

b. ▽▽▽▽▽▽▽▽▽▽▽▽▽ odd / even ______

c.       odd / even ______

Part 4

a. Draw a shape that has 3 angles.

b. Draw a shape that has 5 angles.

c. Draw a shape that has 4 angles.

Copyright © The McGraw-Hill Companies, Inc.

Independent Work

Part 5

a. 69 + 9 = ______

b. 86 + 9 = ______

c. 95 + 9 = ______

d. 9 + 113 = ______

e. 9 + 47 = ______

Lesson 

Name ____________________

Part 6 Write an addition problem and a times problem to figure out the number of squares.

a.

☐
\+ ☐
\+ ☐
\+ ☐
\+ ☐
\+ ☐
______

____

Part 7 Write the sign >, <, or =.

a. 56 ☐ 70 − 20

b. 235 ☐ 5 + 230

c. 30 + 70 ☐ 100

d. 26 ☐ 29 − 4

e. 900 − 300 ☐ 60

f. 5 + 940 ☐ 969

Copyright © The McGraw-Hill Companies, Inc.

Lesson 

Name ______________________________

Part 1 Write the letter or letters for each shape.

C: Circle   Cu: Cube   H: Hexagon   P: Pentagon
Py: Pyramid   Q: Quadrilateral   R: Rectangle   RP: Rectangular Prism
S: Square   Sp: Sphere   T: Triangle

a. 

b. 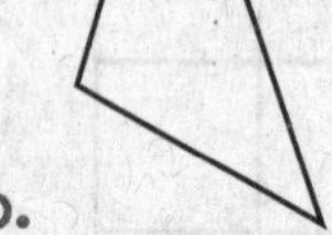

c. 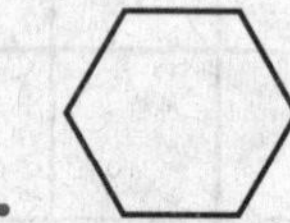

d. 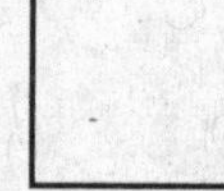

e.

f. 

g. 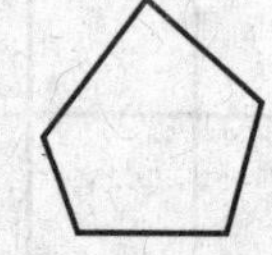

h. 

i. 

j. 

k. 

Part 2

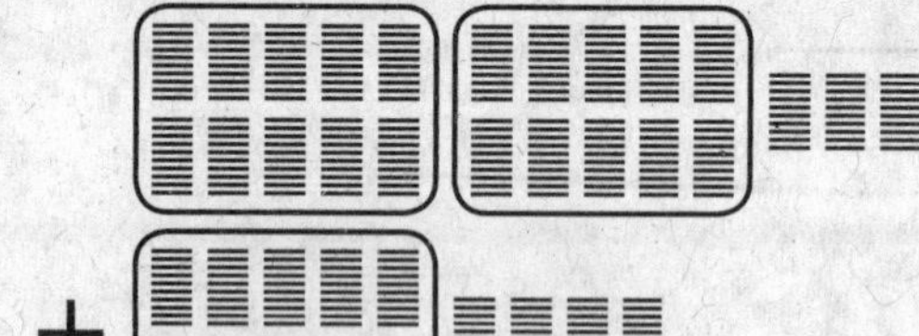

+

a.  + =

Subtract 120

b.  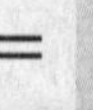 =

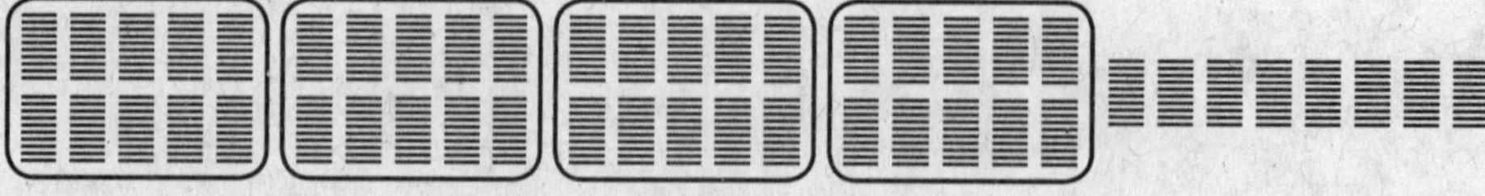

Subtract 350

c.  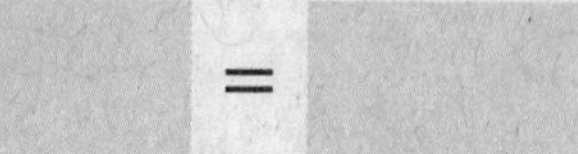 =

Copyright © The McGraw-Hill Companies, Inc.

 Name ____________

Part 3

a. Draw a shape that has 4 angles.

b. Draw a shape that has 6 angles.

c. Draw a shape that has 3 angles.

Part 4

Find the missing number in each family.

a. C —38→ 87 ______

b. 20 —V→ 40 ______

c. 6 —K→ 10 ______

d. 131 —29→ R ______

e. B —56→ 289 ______

f. 16 —40→ T ______

Copyright © The McGraw-Hill Companies, Inc.

Lesson 129

Name ____________________

Part 5

For each row, circle odd or even. Then write the number of objects.

a. odd / even ______

b. odd / even ______

c. odd / even ______

Part 6

a. 3 + 8 = ____	k. 9 + 8 = ____	u. 8 + 8 = ____
b. 6 + 5 = ____	l. 14 − 6 = ____	v. 10 − 6 = ____
c. 7 + 4 = ____	m. 13 − 7 = ____	w. 3 + 7 = ____
d. 12 − 6 = ____	n. 14 − 8 = ____	x. 12 − 5 = ____
e. 8 − 1 = ____	o. 8 + 5 = ____	y. 15 − 8 = ____
f. 16 − 8 = ____	p. 5 + 6 = ____	z. 11 − 5 = ____
g. 4 + 6 = ____	q. 11 − 7 = ____	A. 12 − 8 = ____
h. 6 + 7 = ____	r. 6 + 6 = ____	B. 10 − 7 = ____
i. 10 − 8 = ____	s. 7 + 8 = ____	C. 7 + 6 = ____
j. 7 + 7 = ____	t. 8 + 7 = ____	D. 13 − 8 = ____

Copyright © The McGraw-Hill Companies, Inc.

Part 7

a. 86 + ☐ = ☐

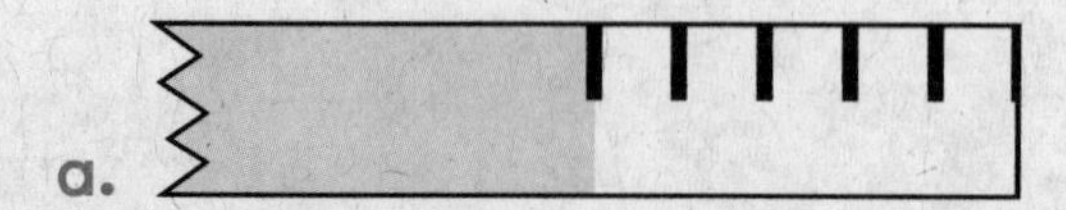

b. ☐ + ☐ = 34

Lesson  Name ____________

Part 1 Work each problem.

a. 120 − 30 = ______ d. 80 − 70 = ______ g. 180 − 90 = ______

b. 70 + 40 = ______ e. 80 + 80 = ______ h. 120 − 20 = ______

c. 130 − 60 = ______ f. 30 + 90 = ______

Part 2 Work each problem.

a. Linda took some money with her. She spent $34. She ended up with $27. How much money did she start out with?

b. There were small rosebushes and large rosebushes in the garden. There were 96 rosebushes in all. 49 bushes were large. How many small bushes were there?

c. There were 35 more well children than sick children. 95 children were well. How many sick children were there?

Part 3 Figure out the number of cents for each row of coins.

a.  ______ cents

b. ______ cents

c. ______ cents

Copyright © The McGraw-Hill Companies, Inc.

Lesson 130

Name ______________________

Part 4

a. 4 + 8 = ____	k. 13 − 8 = ____	u. 8 + 7 = ____
b. 6 + 10 = ____	l. 4 + 3 = ____	v. 6 + 6 = ____
c. 9 + 3 = ____	m. 9 + 7 = ____	w. 11 − 8 = ____
d. 8 + 5 = ____	n. 12 − 7 = ____	x. 14 − 6 = ____
e. 2 + 7 = ____	o. 15 − 10 = ____	y. 3 + 9 = ____
f. 9 + 9 = ____	p. 8 + 6 = ____	z. 20 − 10 = ____
g. 9 − 4 = ____	q. 10 − 9 = ____	A. 7 − 5 = ____
h. 6 + 5 = ____	r. 7 + 5 = ____	B. 12 − 4 = ____
i. 16 − 8 = ____	s. 11 − 2 = ____	C. 10 − 6 = ____
j. 13 − 6 = ____	t. 15 − 7 = ____	D. 7 + 6 = ____

Part 5

Write the time for each clock.

a. ____________

b. ____________

c. ____________

d. ____________

Copyright © The McGraw-Hill Companies, Inc.

Lesson  Name ____________________

Part 6 Write each problem in a column and work it.

a. 126 + 3 + 205 b. 72 + 88 + 120 c. 234 + 31 + 543 d. 184 + 7 + 31

Part 7 Write the sign >, <, or =.

a. 1 yard ☐ 3 feet

b. 1 day ☐ 18 hours

c. 1 week ☐ 7 days

d. 1000 cents ☐ 1 dollar

e. 60 minutes ☐ 1 hour

f. 12 seconds ☐ 1 minute

g. 1 gallon ☐ 5 quarts

h. 12 inches ☐ 1 foot

i. 60 months ☐ 1 year

j. 1 meter ☐ 100 centimeters

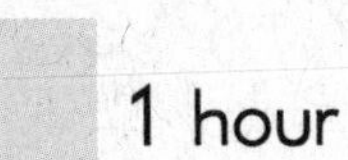

Copyright © The McGraw-Hill Companies, Inc.

Lesson 130

Name ______________________

Part 8 Work each problem.

a. 2 x _____ = 10

b. 5 x 5 = _____

c. 9 x 6 = _____

d. 10 x _____ = 40

e. 4 x _____ = 28

f. 2 x 7 = _____

g. 9 x _____ = 9

h. 5 x _____ = 40

i. 4 x _____ = 20

j. 10 x 3 = _____

k. 2 x 6 = _____

l. 5 x _____ = 30

m. 10 x _____ = 20

n. 5 x 3 = _____

o. 2 x _____ = 8

p. 9 x _____ = 27

Part 9 Work each problem.

a. $\begin{array}{r} \$61.05 \\ +19.81 \\ \hline \end{array}$

b. $\begin{array}{r} \$3.78 \\ -1.18 \\ \hline \end{array}$

c. $\begin{array}{r} \$52.00 \\ -37.00 \\ \hline \end{array}$

d. $\begin{array}{r} \$73.12 \\ 1.18 \\ +13.09 \\ \hline \end{array}$

Copyright © The McGraw-Hill Companies, Inc.

Level C Correlation to Grade 2

Common Core State Standards for Mathematics

Operations and Algebraic Thinking (2.OA)

Represent and solve problems involving addition and subtraction.

1. Use addition and subtraction within 100 to solve one- and two-step word problems involving situations of adding to, taking from, putting together, taking apart, and comparing, with unknowns in all positions, e.g., by using drawings and equations with a symbol for the unknown number to represent the problem.

Lessons	**WB 1: 33–41, 65** **WB 2: 93, 94, 107, 121–125, 127, 128, 130** **TB: 42–100, 102, 103, 105–112, 114–116, 118–120, 123, 129**

Operations and Algebraic Thinking (2.OA)

Add and subtract within 20.

2. Fluently add and subtract within 20 using mental strategies. By end of Grade 2, know from memory all sums of two one-digit numbers.

Lessons	**WB 1: 1–39, 41–70** **WB 2: 71–80, 82, 85–122, 129, 130** **TB:42–49, 56, 58, 59, 62–64, 68–73, 75, 77, 78, 81–83, 85–87, 95, 96, 99–106, 110–115, 119, 121–127**

Operations and Algebraic Thinking (2.OA)

Work with equal groups of objects to gain foundations for multiplication.

3. Determine whether a group of objects (up to 20) has an odd or even number of members, e.g., by pairing objects or counting them by 2s; write an equation to express an even number as a sum of two equal addends.

Lessons	**WB 1: 55–60** **WB 2: 128, 129**

Operations and Algebraic Thinking (2.OA)

Work with equal groups of objects to gain foundations for multiplication.

4. Use addition to find the total number of objects arranged in rectangular arrays with up to 5 rows and up to 5 columns; write an equation to express the total as a sum of equal addends.

Lessons	**WB 1: 47, 48** **WB 2: 71, 73** **TB: 45–47, 49, 75**

Copyright © The McGraw-Hill Companies, Inc.

Number and Operations in Base Ten (2.NBT)

Understand place value.

1. Understand that the three digits of a three-digit number represent amounts of hundreds, tens, and ones; e.g., 706 equals 7 hundreds, 0 tens, and 6 ones. Understand the following as special cases:
   a. 100 can be thought of as a bundle of ten tens — called a "hundred."
   b. The numbers 100, 200, 300, 400, 500, 600, 700, 800, 900 refer to one, two, three, four, five, six, seven, eight, or nine hundreds (and 0 tens and 0 ones).

Lessons	**WB 1: 1–4, 7, 10–25, 27, 29, 36, 59, 68** **WB 2: 74, 116, 118, 126–129** **Student Practice Software: Block 1 Activities 1 and 2, Block 4 Activities 2 and 3, Block 6 Activity 6**

Number and Operations in Base Ten (2.NBT)

Understand place value.

2. Count within 1000; skip-count by ***2,** 5s, 10s, and 100s.

Lessons	**WB 1: 11, 12, 15–29, 31–60, 63–65, 67–69** **WB 2: 73–77, 80, 88, 90–94, 98, 101, 102, 116–120, 122–126, 130** **TB: 41, 62, 63, 65, 68, 69, 72, 73, 76–80, 86, 87, 89 ,90, 95–117, 119, 120, 122, 125–127** **Student Practice Software: Block 3 Activity 2**

*Denotes California-only content.

Number and Operations in Base Ten (2.NBT)

Understand place value.

3. Read and write numbers to 1000 using base-ten numerals, number names, and expanded form.

Lessons	**WB 1: 1–7, 11–37, 39, 40, 42, 43, 45–51, 54, 58, 59, 68** **WB 2: 74** **TB: 82, 118** **Student Practice Software: Block 2 Activity 3, Block 4 Activity 4, Block 5 Activity 4**

Number and Operations in Base Ten (2.NBT)

Understand place value.

4. Compare two three-digit numbers based on meanings of the hundreds, tens, and ones digits, using >, =, and < symbols to record the results of comparisons.

Lessons	**WB 1: 31, 36** **WB 2: 112, 113, 115, 119–121, 124–128** **Student Practice Software: Block 2 Activity 5**

Copyright © The McGraw-Hill Companies, Inc.

Number and Operations in Base Ten (2.NBT)

Use place value understanding and properties of operations to add and subtract.

5. Fluently add and subtract within 100 using strategies based on place value, properties of operations, and/or the relationship between addition and subtraction.

Lessons	WB 1: 1–70 WB 2: 71–91, 93–130 TB: 47–50, 52–91, 94–96, 98–100, 102–115, 117–129 Student Practice Software: Block 3 Activities 1, 3, 4, 5; Block 5 Activity 2

Number and Operations in Base Ten (2.NBT)

Use place value understanding and properties of operations to add and subtract.

6. Add up to four two-digit numbers using strategies based on place value and properties of operations.

Lessons	WB 1: 4–10, 12, 16–20, 22–56, 60–70 WB 2: 71–94, 96–98, 100–102, 110, 111, 113, 122, 127–130 TB: 46, 47, 49, 50, 52–55, 57, 59–66, 69–71, 73, 75–82, 84, 86, 88, 89, 92–100, 102, 107–114, 117, 119, 122, 124, 125, 126, 128 Student Practice Software: Block 1 Activity 3

Number and Operations in Base Ten (2.NBT)

Use place value understanding and properties of operations to add and subtract.

7. Add and subtract within 1000, using concrete models or drawings and strategies based on place value, properties of operations, and/or the relationship between addition and subtraction; relate the strategy to a written method. Understand that in adding or subtracting three-digit numbers, one adds or subtracts hundreds and hundreds, tens and tens, ones and ones; and sometimes it is necessary to compose or decompose tens or hundreds.

Lessons	WB 1: 5–8, 10–70 WB 2: 71, 73–89, 92–95, 100, 106, 107, 109, 111–115, 119–121, 124, 125, 127–130 TB: 46–50, 62–69, 71, 73–83, 85, 86, 88–117, 119–123, 125, 127, 128 Student Practice Software: Block 1 Activity 5, Block 2 Activity 2, Block 5 Activity 5

Number and Operations in Base Ten (2.NBT)

Use place value understanding and properties of operations to add and subtract.

***7.1 Use estimation strategies to make reasonable estimates in problem solving.**

Lessons	WB 2: 75–84, 88, 90–94, 97, 98, 100, 102 TB: 76–79, 95, 96, 98, 99, 105, 107–109, 111–114, 125, 126, 128

*Denotes California-only content.

Copyright © The McGraw-Hill Companies, Inc.

Number and Operations in Base Ten (2.NBT)

Use place value understanding and properties of operations to add and subtract.

8. Mentally add 10 or 100 to a given number 100–900, and mentally subtract 10 or 100 from a given number 100–900.

Lessons	WB 1: 22–24, 42, 43, 50, 58, 59 WB 2: 75–78, 80, 89, 112, 117–121 TB: 89

Number and Operations in Base Ten (2.NBT)

Use place value understanding and properties of operations to add and subtract.

9. Explain why addition and subtraction strategies work, using place value and the properties of operations.

Lessons	WB 2: 115–120

Measurement and Data (2.MD)

Measure and estimate lengths in standard units.

1. Measure the length of an object by selecting and using appropriate tools such as rulers, yardsticks, meter sticks, and measuring tapes.

Lessons	WB 1: 30–41, 43, 44, 46, 49, 54, 59–68 WB 2: 115–120, 122, 124–127

Measurement and Data (2.MD)

Measure and estimate lengths in standard units.

2. Measure the length of an object twice, using length units of different lengths for the two measurements; describe how the two measurements relate to the size of the unit chosen.

Lessons	WB 1: 34, 35, 41, 43, 44, 46, 49, 51, 53, 54 WB 2: 85–87, 116

Measurement and Data (2.MD)

Measure and estimate lengths in standard units.

3. Estimate lengths using units of inches, feet, centimeters, and meters.

Lessons	WB 2: 85–87 TB: 125–127

Copyright © The McGraw-Hill Companies, Inc.

Measurement and Data (2.MD)

Measure and estimate lengths in standard units.

4. Measure to determine how much longer one object is than another, expressing the length difference in terms of a standard length unit.

Lessons	WB 2: 115–120, 122

Measurement and Data (2.MD)

Relate addition and subtraction to length.

5. Use addition and subtraction within 100 to solve word problems involving lengths that are given in the same units, e.g., by using drawings (such as drawings of rulers) and equations with a symbol for the unknown number to represent the problem.

Lessons	TB: 52, 53, 57–71, 74, 75, 83, 84, 94, 96, 98, 123, 126

Measurement and Data (2.MD)

Relate addition and subtraction to length.

6. Represent whole numbers as lengths from 0 on a number line diagram with equally spaced points corresponding to the numbers 0, 1, 2, …, and represent whole-number sums and differences within 100 on a number line diagram.

Lessons	WB 1: 51–54, 56, 59, 67–70 WB 2: 125, 127, 128 Student Practice Software: Block 6 Activity 1

Measurement and Data (2.MD)

Work with time and money.

7. Tell and write time from analog and digital clocks to the nearest five minutes, using a.m. and p.m. ***Know relationships of time (e.g., minutes in an hour, days in a month, weeks in a year).**

Lessons	WB 2: 91–107, 109, 112, 114, 116, 118, 120, 122–124, 130 TB: 85–88, 90, 98, 99, 101, 103, 106, 108, 109, 112, 115, 117–119, 121–123, 126 Student Practice Software: Block 6 Activity 2

*Denotes California-only content.

Copyright © The McGraw-Hill Companies, Inc.

Measurement and Data (2.MD)

Work with time and money

8. Solve word problems involving dollar bills, quarters, dimes, nickels, and pennies, using $ and ¢ symbols appropriately. *Example: If you have 2 dimes and 3 pennies, how many cents do you have?*

Lessons	WB 2: 94, 130 TB: 43, 46, 48, 54–56, 58, 62, 64, 65, 73, 75, 76, 80, 86, 87, 90, 93, 97–116, 118–121, 123, 124, 126, 128 Student Practice Software: Block 4 Activity 3, Block 5 Activity 6

Measurement and Data (2.MD)

Represent and interpret data.

9. Generate measurement data by measuring lengths of several objects to the nearest whole unit, or by making repeated measurements of the same object. Show the measurements by making a line plot, where the horizontal scale is marked off in whole-number units.

Lessons	WB 2: 124–127 Student Practice Software: Block 6 Activity 3

Measurement and Data (2.MD)

Represent and interpret data.

10. Draw a picture graph and a bar graph (with single-unit scale) to represent a data set with up to four categories. Solve simple put-together, take-apart, and compare problems using information presented in a bar graph.

Lessons	WB 2: 117–120, 122, 123, 125 TB: 121, 124, 126, 127 Student Practice Software: Block 6 Activity 4

Geometry (2.G)

Reason with shapes and their attributes.

1. Recognize and draw shapes having specified attributes, such as a given number of angles or a given number of equal faces. Identify triangles, quadrilaterals, pentagons, hexagons, and cubes.

Lessons	WB 1: 43–49, 54, 56–58 WB 2: 111–118, 127–129 TB: 50, 51, 53, 55–57, 59, 60, 66, 68, 70, 74, 76, 79, 84, 110 Student Practice Software: Block 4 Activity 5

Copyright © The McGraw-Hill Companies, Inc.

Geometry (2.G)

Reason with shapes and their attributes.

2. Partition a rectangle into rows and columns of same-size squares and count to find the total number of them.

Lessons	WB 1: 47, 48 WB 2: 121–126, 128 TB: 45, 46, 48, 49, 126 Student Practice Software: Block 4 Activity 6

Geometry (2.G)

Reason with shapes and their attributes.

3. Partition circles and rectangles into two, three, or four equal shares, describe the shares using the words *halves, thirds, half of, a third of,* etc., and describe the whole as two halves, three thirds, four fourths. Recognize that equal shares of identical wholes need not have the same shape.

Lessons	WB 2: 125–127 Student Practice Software: Block 6 Activity 5

Copyright © The McGraw-Hill Companies, Inc.

Standards for Mathematical Practice

Connecting Math Concepts addresses all of the Standards for Mathematical Practice throughout the program. What follows are examples of how individual standards are addressed in this level.

1. Make sense of problems and persevere in solving them.

Word Problems (Lessons 12–25, 28–77): Students learn to identify specific types of word problems (i.e., start-end, comparison, classification) and set up and solve the problems based on the specific problem types.

2. Reason abstractly and quantitatively.

Addition/Subtraction (Lessons 1–64, 73–115): Beginning in Lesson 45, students count objects in two groups, write the number for each group, and then add them to find the total objects. They connect the written numbers with quantities while learning the concept of addition.

3. Construct viable arguments and critique the reasoning of others.

Estimation (Lessons 65–94): Students learn how to round numbers and then apply that knowledge to word problems involving estimation. They work the original problem and the estimation problem and then compare answers to verify that the estimated answer is close to the exact answer. Students can construct an argument to persuade someone whether an estimated answer is reasonable.

4. Model with mathematics.

Number Families (Lessons 1–29): Students learn to represent three related numbers in a number family. Later, they apply number families to model and solve word problems.

5. Use appropriate tools strategically.

Throughout the program (Lessons 1–130) students use pencils, workbooks, lined paper, and textbooks to complete their work. They use rulers to measure lines. They use the computer to access the Practice Software where they apply the skills they learn in the lessons.

6. Attend to precision.

Measurement (Lessons 30–43, 59–78, 95–103): When measuring lines and finding perimeter and area, students learn to include the correct unit in the verbal and written answers. They also include units in answers to word problems that involve specific units.

7. Look for and make use of structure.

Mental Math (Lesson 9 and frequently throughout): Students build computational fluency by learning patterns inherent in different problem types, such as +/– 10 and +/– 100.

8. Look for and express regularity in repeated reasoning.

Multiplication (Lessons 32–49): Students are introduced to the concept of multiplication by thinking of it as repeated counting. For example, 2 x 5 tells us to count by 2 five times.

Copyright © The McGraw-Hill Companies, Inc.